D1191252

ECOLOGICAL TOXICITY TESTING

Scale, Complexity, and Relevance

Edited by
John Cairns, Jr.

University Distinguished Professor
Department of Biology
and
Director
University Center for Environmental and Hazardous Materials Studies
Virginia Polytechnic Institute and State Univerity
Blacksburg, Virginia

B. R. Niederlehner
Laboratory Specialist and Staff Scientist
University Center for Environmental and Hazardous Materials Studies
Virginia Polytechnic Institute and State Univerity
Blacksburg, Virginia

LEWIS PUBLISHERS
Boca Raton   Ann Arbor   London   Tokyo

Library of Congress Cataloging-in-Publication Data

Ecological toxicity testing : scale, complexity, and relevance /
  edited by John Cairns, Jr. and B. R. Niederlehner.
    p.   cm.
  Includes bibliographical references and index.
  ISBN 0-87371-599-3
    1. Toxicity testing.   2. Pollution--Environmental aspects.
  I. Cairns, John, 1923– .   II. Niederlehner, B. R.
  QH541.15.T68E24   1994
  574.2′5--dc20   94-19832
                                                                          CIP

© 1995 by CRC Press, Inc.
Lewis Publishers is an imprint of CRC Press

No claim to original U.S. Government works
International Standard Book Number 0-87371-599-3
Library of Congress Card Number 94-19832
Printed in the United States of America 1 2 3 4 5 6 7 8 9 0
Printed on acid-free paper

Preface

In one respect, we think this book is ideally timed. Much has been accomplished in the development and validation of predictive methods for estimating the ecological consequences of toxic chemicals. Since 1985, the year in which several books on a similar topic came out, standard microcosm methods have become available; mesocosms and outdoor microcosms have been used repeatedly for evaluations of pesticides; and both single species and multispecies toxicity tests have been subjected to critical evaluations of their predictive accuracy. It will be clear, from both Chapter 1 and the summary chapter (13), that we do not believe that the evolutionary process in ecotoxicological methods is complete, but much has been accomplished.

However, in another important respect, this book is badly timed. Almost 30% of our scheduled contributors found that circumstances prevented them from meeting their obligations. We think this is almost certainly related to the severe budget reductions that higher education is now experiencing and the resulting pressures on professional time. While we would have liked to fill the resulting gaps in (a) extrapolations and validation, (b) terrestrial systems, (c) field enclosures, and (d) whole lake manipulations, the delay would have dated the contributions that were completed on schedule. This seemed unfair to those colleagues who provided their manuscripts on time. We and the reviewers felt that a critical mass of information had been achieved and that it provides an overview of the different scales at which problems of toxic chemicals can be approached.

The Editors

John Cairns, Jr., Ph.D., is University Distinguished Professor of Environmental Biology in the Department of Biology and Director of the University Center for Environmental and Hazardous Materials Studies at Virginia Polytechnic Institute and State University, Blacksburg. He received his A.B. from Swarthmore College and his M.S. and Ph.D. from the University of Pennsylvania. He completed a postdoctoral course in isotope methodology at Hahnemann Medical College, Philadelphia. He was Curator of Limnology at the Academy of Natural Sciences of Philadelphia for 18 years, and has taught at various universities and field stations. Professional certifications include Certified Environmental Professional by the National Association of Environmental Professionals, Qualified Fishery Administrator by the American Fisheries Society, and Senior Ecologist by the Ecological Society of America.

Among Dr. Cairns' awards are Member, National Academy of Sciences; Fellow, American Academy of Arts and Sciences; Fellow, American Association for the Advancement of Science; Foreign Member, Linnean Society of London; Founder's Award of the Society for Environmental Toxicology and Chemistry; United Nations Environmental Programme Medal; U.S. Presidential Commendation for Environmental Activities; Icko Iben Award from the American Water Resources Association; B. Y. Morrison Medal, awarded at the Pacific Rim Conference of the American Chemical Society; Superior Achievement Award, U.S. Environmental Protection Agency; Charles B. Dudley Award for excellence in publications from the American Society for Testing and Materials; Life Achievement Award in Science from the Commonwealth of Virginia and the Science Museum of Virginia; and American Fisheries Society Award of Excellence. Dr. Cairns has served as both vice president and president of the American Microscopical Society; has served on 17 National Research Council committees, two as chair; is presently serving on 14 editorial boards; and has served on the Science Advisory Board of the International Joint Commission (U.S. and Canada) and on the U.S. EPA Science Advisory Board.

Approximately one-fourth of his nearly 1200 publications are sole-authored. One of the most recent of his 48 books is *Restoration of Aquatic Ecosystems: Science, Technology, and Public Policy* (Committee Chair), National Academy Press, 1992.

B. R. Niederlehner, M.S., is Laboratory Specialist and Staff Scientist in the University Center for Environmental and Hazardous Materials Studies at Virginia Polytechnic Institute and State University. She has an M.S. degree in Biology (aquatic toxicology) and 19 years experience in assessing the effects of stress on populations, communities, and ecosystems. Niederlehner is an author of over 30 articles within the areas of risk assessment and microbial community ecology.

Acknowledgments

We are deeply indebted to Darla Donald, without whose skilled editorial assistance this book would not be possible, and to Teresa Moody for transcribing the voluminous correspondence always involved in the editing of a book. We are most appreciative of the reviewers whose opinions and specialized knowledge raised the level of the contributions. Some persons reviewed one or more chapters, a crucial activity for all scientific publications but one for which, regrettably, too little credit is given by academic institutions. Names of reviewers follow in alphabetical order: Scott E. Belanger, Procter & Gamble Company; William H. Clements, Colorado State University; John Harte, University of California, Berkeley; Alan G. Heath, Virginia Tech; Guy R. Lanza, East Tennessee State University; John R. Lauth, Virginia Tech; Paul V. McCormick, South Florida Water Management District; Kenneth T. Perez, United States Environmental Protection Agency; Kurt W. Pontasch, University of Northern Iowa; John B. Sprague, J. B. Sprague Associates; and William H. Waller, University of North Texas.

Contributors

John Cairns, Jr.
University Center for Environmental
  and Hazardous Materials Studies
Virginia Polytechnic Institute and
  State University
Blacksburg, Virginia 24061

Kenneth L. Dickson
Institute of Applied Sciences
University of North Texas
P.O. Box 13078
Denton, Texas 76203

Lenore Fahrig
Department of Biology
Carleton University
Ottawa, Ontario K1S 5B6
Canada

Kathryn Freemark
National Wildlife Research Centre
Environment Canada
c/o U. S. Environmental Protection
  Agency
Environmental Research Laboratory
Corvallis, Oregon 97333

Willem Halffman
Science Dynamics
University of Amsterdam
Nieuwe Achtergracht 166
1018 WV Amsterdam
The Netherlands

Philip C. Johnson
U.S. Environmental Protection
  Agency
Region VIII
999 18th Street, Suite 500
Denver, Colorado 80202

Zane B. Johnson
Department of Biological Science
University of North Texas
P.O. Box 5218
Denton, Texas 76203

James H. Kennedy
Institute of Applied Sciences
University of North Texas
P.O. Box 13078
Denton, Texas 76203

Donald I. Mount
Director of Aquatic Toxicity
  Programs
AScI Corporation
112 East Second Street
Duluth, Minnesota 55805

B. R. Niederlehner
University Center for Environmental
  and Hazardous Materials Studies
Virginia Polytechnic Institute and
  State University
Blacksburg, Virginia 24061

Benjamin R. Parkhurst
The Cadmus Group, Inc.
P. O. Box 546
309 South Fourth, Suite 202
Laramie, Wyoming 82070

Kenneth T. Perez
U. S. Environmental Protection
  Agency
Environmental Research Laboratory
27 Tarzwell Drive
Narragansett, Rhode Island 02882

Kurt W. Pontasch
Department of Biology
University of Northern Iowa
Cedar Falls, Iowa 50614–0421

Eric P. Smith
Department of Statistics
Virginia Polytechnic Institute and
  State University
Blacksburg, Virginia 24061

Glenn W. Suter II
Environmental Sciences Division
Oak Ridge National Laboratory
P.O. Box 2008, Mail Stop 6038
Oak Ridge, Tennessee 37831

Dedication

This book is dedicated to Robert K. Enders, who was my advisor during my undergraduate years at Swarthmore College. Bob Enders' understanding and help launched my academic career. As happened with many people in the early 1940s, my undergraduate education, which started at Pennsylvania State University, was interrupted by World War II. During the war, I acquired a spouse and a child. After being released from the Navy, I could only find affordable housing at my mother-in-law's in suburban Philadelphia. Penn State was willing to readmit me, but it had no family housing at the time. I took a compass and, using my mother-in-law's house as a center, drew a circle with a 50-mile radius and began to apply to all colleges and universities within the circle. Most were already overcrowded and, since my grades for the first two years were not exemplary, I was rejected by all of them. Finally, I extended my area to include Swarthmore, not knowing its prominent reputation and how difficult it was to get admitted, even in those days. I still remember my interview with the dean, who asked me how I selected Swarthmore. He then had a coughing fit, which I later realized was an instinctive attempt to cover side-splitting laughter. However, when he wiped the tears from his eyes, we continued the interview. I was then sent to Bob Enders' office because I had indicated an interest in a biology degree. The interview with Enders is as vivid today as the day it happened, although I now fully realize my indebtedness to him. Just before my graduation from Swarthmore, Enders had another low key conversation with me, not on whether I intended to enroll in graduate school but where! At some point in the conversation, Dr. Enders pulled from his pocket a slip of paper with an appointment time and date already indicated with Professor David H. Wenrich in the Biology Department of the University of Pennsylvania. I already owed so much to Dr. Enders that I thought the least I could do was attend graduate school until I was thrown out, but, astonishingly, I acquired an M.S. and Ph.D. Both David Wenrich and Ruth Patrick have had notable influences on my career after my graduation from Swarthmore; however, the necessary preliminary, without which further academic associations would have been impossible, was Dr. Enders. He continued to influence my career until 1984, which is the last time we saw each other. Frequently, students appear in my office who I have never seen before and for whom I have no formal advising responsibility. What they are after is advice on career development, which usually takes at least an hour, and not uncommonly two hours. This is, in a sense, partial payment of my great debt to Bob Enders, and I only hope that I am occasionally able to help a student as much as he helped me.

Contents

CHAPTER 1

The Genesis of Ecotoxicology

John Cairns, Jr.

INTRODUCTION

A few years after World War II, which focused nearly all of the U.S.'s energy on warfare, concern increased about the ecological condition of natural systems, particularly those receiving industrial waste discharges. Interest in environmental toxicology (more recently called ecotoxicology) was generated by a practical and immediate concern: determining whether a particular waste discharge was harming the environment. This impetus came not from the scientific community but from society as a whole. In response, tests were developed in industrial laboratories in conjunction with activities of state and federal regulatory agencies. The first provisional standard method for toxicity testing of aquatic organisms was developed in the research laboratories of the Atlantic Refining Company by Harte, Doudoroff, and Greenbank.[1] This test subsequently reached the status of a standard method through the formal system of the American Society for Testing and Materials. As a consequence of these practical origins, those carrying out toxicity tests were not immediately concerned with large-scale prediction of environmental outcome but rather with the need to generate information for decisionmaking.

Toxicological effects occur at many scales: at the cellular level, enzymes are induced; at the population level, there is mortality and selection; at the community level, there is replacement of species. Studying effects at the landscape level is more difficult; however, increasing evidence indicates that landscapes and even larger scales are the inherently correct scale on which to understand some environmental problems. For example, chemical stressors, even those emitted from a point source discharge, may spread over a sizable area through a variety of transportation mechanisms. Kern[2] postulated that the

0-87371-599-3/95/$0.00+$.50

haze in the Arctic region results from industrial smog spreading from Europe. Additionally, Hirao and Patterson[3] furnish persuasive evidence that the damage from car exhaust originating in the coastal California cities is detectable in the Sierra Nevada Mountains located in the eastern part of the state. An even more compelling case for landscape effects is the evidence provided by Spencer and Cliath,[4] who have shown that over 90% of a number of toxic pesticides had volatilized into the atmosphere within a week after application, presumably to then be distributed in such a way that they would affect areas quite distant from the area of application. Kurtz and Atlas[5] have shown that volatilization and atmospheric transport of hexachlorocyclohexanes have resulted in the contamination of surface water throughout the Pacific Ocean. In addition, persuasive evidence shows that small-scale environmental impacts can become landscape stresses when similar impacts are repeated over a large area.[6-8] Since more attention is being given to landscape-level phenomena by research investigators, more and more instances of environmental problems on a large spatial scale are being described.

American movie audiences have realized that toxicants are spread over landscapes ever since Cary Grant was threatened by a crop-duster plane on a lonely road in the American Midwest in the Alfred Hitchcock film *North by Northwest,* released in 1956. By that time, single species toxicity tests with fish and other aquatic organisms were already in use. Surely, the field of environmental toxicology could have developed a means of looking at larger scales at that point. Instead, many environmental toxicologists (myself included) were then engrossed in determining the effects of chemical/physical characteristics of the dilution water and the ways in which these characteristics mediated the toxic response. In hindsight, questions about the modifying effects of environmental variables were answered to a degree far out of proportion to equally important questions of modifying effects of measuring response at different levels of biological and ecological organization.

Perhaps this focus on the individual and population levels can be traced to the incomplete separation of environmental toxicology from the field of human health effects, where methods and procedures in mammalian toxicology were already moderately well developed. While the object of protection in the later case is a clearly a single species, *Homo sapiens*, the object of protection in environmental toxicology is much broader and harder to delimit.

Students of ecotoxicology are sometimes incredulous when they discover that questions about the applicability of basic toxicity test methods to prediction of environmental outcome on a larger scale were not asked sooner. For a time, investigators were not interested in assessing the robustness of predictions of response based on single species laboratory tests in more complex natural systems. Instead, the ability to replicate laboratory results was used as evidence of their utility. That is, if a "round robin" was carried out with the same species (preferably a pedigreed strain), with a standard test material under standardized laboratory conditions at different testing organizations, and the results corresponded quite well or were reasonably comparable, the *method* was considered validated. If the method provided repeatable response thresholds, then the assumption was made that, by testing "the most sensitive species" in a system, the system would inevitably be protected because the response thresholds of all other species were higher.[9] I have previously pointed out a number of weaknesses in this position:[10]

1. How could we be sure that the limited array of test species (well under 100) actually included the most sensitive of the literally millions of species in natural systems?
2. Even if the laboratory test species were, in fact, the most sensitive to a particular chemical, ample evidence in the literature shows that the same species would not be the most sensitive to all chemicals.
3. The species might respond differently when not isolated from other species, and chemicals might act differently when in mixtures.

In addition, while single species laboratory tests might be relatively inexpensive to conduct compared to more inclusive tests, this cost savings disappears when the cost of bad management decisions resulting from inaccurate predictions is considered. In addition, although microcosm and mesocosm test systems have been thought to be far too variable to produce consistent results, experimental evidence is lacking to support this conclusion.

To modify an old folk tale, while the emperor was not entirely without clothes, he certainly needed to be wearing a few more. It was, and is, a large jump to assume that a fish in a jar or a rat in a cage can accurately anticipate the response of complex ecosystems possessing attributes and variables not even considered in the laboratory tests. However, the persistent appeal of simple and standardized tests could be attributed to a number of factors:

1. The high degree of specialization in science tended to focus on reductionist methods (a particular component of the system), and integrative science (interactions of the components) had not, until more recently, received much attention or appreciation.
2. Irreconcilable design differences occur between tests that are easily repeatable and tests that are more representative of field response. The standardization of toxicological tests led to the elimination of the very variables that contributed to ecological realism.
3. By the time a test was in general use, considerable capital investment and inertia were involved. The field of environmental toxicology was beginning to be accepted. Pointing out discrepancies between the capabilities of test methods and desirable levels of performance could lead to a loss in confidence, consulting firm work, and regulatory authority. Few ecotoxicologists would wish to precipitate a diversion of resources from testing, any testing, to legal challenges.

Discussions of each of these possibilities follows.

STUDIES OF FRAGMENTS RATHER THAN SYSTEMS

Basically, the scientific method involves randomizing or controlling all variables in a system except for the one being intensively studied (i.e., the reductionist approach). This approach has been enormously successful. Reductionist science has been responsible for breakthroughs of enormous importance and will undoubtedly continue to elucidate many new pivotal scientific concepts in the future. One need not denigrate reductionist science to espouse the value of integrative science in understanding how components collectively interact in systems.[11] Both approaches are valuable.

Ecotoxicology implies the interfacing of the fields of ecology and environmental toxicology. I have previously suggested that the term **ecotoxicologist** could describe an individual who uses ecological parameters to assess toxicity.[12] If this definition is accepted, ecotoxicology would be the result of an interface between the fields of ecology and toxicology. Further, a more restrictive definition of ecotoxicologist would be "one who assesses the effects of toxic substances on ecosystems."[12] The latter seems a more realistic definition because one goal of most environmental management is to protect entire ecosystems, not just isolated components thereof.

STANDARDIZATION AND THE ELIMINATION OF ECOLOGICAL ATTRIBUTES

Uniform, standard tests are essential as a means of comparing various toxicants before they are released into ecosystems. Estimating relative hazard provides information on whether compound A is more or less toxic to a particular species than compound B. Whether the effects are acute or chronic can also be determined. With approximately 76,000 chemicals in daily use and choices of chemicals for any particular application, few people would dispute the need for a standard unit of measure. However, the standard unit of measure is not necessarily a good means of predicting actual environmental outcome in ecosystems that vary in many significant attributes from one area to another. Thus, the standard unit of measure can provide much information useful for priority setting, selecting the least toxic among an array of chemicals, determining the capabilities for certifying a toxicological laboratory, and determining where scarce resources should be used to reduce hazard. However, a test suitable for these purposes is almost certainly not going to be the best available for predicting responses of complex multivariate systems, which differ markedly from one geographic region to another and for which endpoints are at a different level of biological organization. It would be most convenient if tests suitable for the ranking hazards would also be suitable for predicting responses at the community and ecosystem levels. However, no robust scientific evidence supports this.

The U.S. Environmental Protection Agency (U.S. EPA) led the way in standardizing environmental toxicity tests because standardization might provide a fair and unbiased approach to regulation, reducing the need for, and challenges to, professional judgment. In fact, while simplified, standardized tests reduce the need for professional judgment at the testing stage, they require more professional judgment at the stage of predicting actual environmental outcome from the results of those tests. In contrast, multispecies microcosm and mesocosm toxicity tests require more professional judgment at the testing stage but make it possible to use the same endpoints in the test as are used to evaluate the health of natural systems. This reduces the validation problem significantly.

The relationship between classical ecologists and environmental toxicologists has never been a strong one, and an uncharitable person might well describe it as tenuous. Classical ecologists tend to visit pristine ecosystems, such as the Galapagos Islands, whereas environmental toxicologists are primarily in the laboratory for their professional activities. Fortunately, in the last few years, associations have increased in number and frequency. Further cooperation will facilitate the validation of predictive models based on laboratory evidence in natural systems. The potentials for productive interactions between the two groups are discussed by Cairns.[13] Like all relationships, this one will not flourish without care and attention, which will inevitably divert a significant amount of energy from one's area of specialization. Unless the circumstances change in the academic world as well as in the larger professional world, status will inevitably be lost within the area of specialization. This is indeed unfortunate because society needs better predictors of the effects of anthropogenic stresses upon ecosystems.

The drive toward standardization led to reducing variability in the test container in both physical and chemical characteristics (e.g., temperature and dissolved oxygen concentration) and in standardizing the test organisms themselves. For example, instead of having all life history stages in a single test container, most commonly a single life history stage was present for a particular species. Furthermore, feeding conditions were standardized, although they rarely are in natural systems; attempts were made to eliminate disease, although the number of individuals in natural populations succumbing to disease is substantial; predator/prey interactions did not exist; competitive interactions with other species were eliminated; and, possibly most importantly, species difficult to culture in the laboratory are not given serious consideration.

Most of the endpoints used in standardized, single species toxicity tests are based on no-observable-adverse effects on such endpoints as lethality, growth and reproductive success, and, more rarely, such things as locomotion and the like. More recently, however, emphasis is on determining ecosystem health of whole natural systems. As in human medicine, the focus is shifting from eliminating symptoms of the disease and other malfunction to focusing on robust health and condition. Similarly, the no-observable-adverse effects concentration or level is, at least as an aspiration, being replaced with an emphasis on ecosystem health and condition. How these will be measured is not yet entirely clear. However, what is abundantly clear is the fact that ecosystem health is extraordinarily difficult to measure in a standardized test that eliminates most of the attributes of natural systems, including natural variability.

Although standardized tests will undoubtedly persist for the reasons given earlier, the trend toward testing complex, multivariate systems is too well established and has too much momentum to disappear. Additionally, the focus on ecosystem health ensures a closer linkage between the results of laboratory toxicity tests and the well being of ecosystems. Finally, the attention to condition or health inevitably means that a new set of endpoints or parameters to assess ecosystem stress will appear. Since many of these will be more highly variable than the standardized toxicity tests, a quite different set of statistical methods will be used to assess their soundness. What is not clear is whether these future developments will involve the interfacing of classical ecology and environmental toxicology or whether the new set of journals (e.g., *Ecotoxicology* and the *Journal of Aquatic Ecosystem Health*) will mean that, instead of an interface, new subprofessions will appear. Since it is difficult to change existing professional societies with the rapidity that science is changing, the latter seems most probable.

ECONOMIC FACTORS

Toxicity testing is a significant industry. It is, moreover, a service industry where clients may not see the tests at all or, at most, a small fraction of the testing process. Therefore, credibility is important in two ways: (1) the credibility of the practitioners carrying out the tests and (2) the credibility of the tests themselves. Developing the capability to carry out such tests means a significant capital investment with a concomitant significant investment in the personnel responsible for running the tests. Every change in procedures and methodology is costly. Furthermore, probably 98 or 99% of the people who are professional toxicity testers are not primarily research investigators. A significant problem exists because scientists must be intolerant of uncertainty. If there are weaknesses in environmental toxicity testing, the science of ecotoxicology should identify and address them. However, pointing out weaknesses can undermine confidence in the tests used for day-to-day regulatory decisions that must proceed despite uncertainty. A lack of confidence in regulatory tools may result in significant financial losses due to retraining and retooling. This was impressed on me rather forcefully when I received the Founders' Award from the Society for Environmental Toxicology and Chemistry in 1982. The recipient of this award is supposed to give a plenary session address in association with the presentation of the award. The title of my address was "Are Single

Species Toxicity Tests Alone Adequate for Estimating Environmental Hazard?" and was subsequently published in *Hydrobiologia*.[14] A significant number of people commented during the question period and afterwards that the reasoning appeared sound, and this was just another phase in the development of environmental toxicology testing. However, a much larger number viewed it as an out-and-out attack on single species toxicity testing and, therefore, an attack on the practices of most of the people in the field. There were, in the address itself and in subsequent publications, frequent caveats to the effect that single species tests were the only means of determining lethality, growth and reproductive success, respiratory efficiency, and a variety of other toxicological endpoints characteristic of single species tests. In addition, another important caveat was that single species testing is best for screening and rangefinding to determine priorities, to compare relative toxicity of chemicals and the like. My intent was not to abolish single species testing but to expand the array of toxicity tests available, particularly by adding tests at higher levels of biological organization. However, I naively assumed that this was entirely an academic matter and was not as aware as I should have been that this was viewed as an attack on a sizable, established business and on the livelihoods of a number of people. Many went so far to state that they agreed with me in principle but that confidence in toxicity testing was just beginning to be widely established, and I should not endanger the further establishment of this confidence by criticizing the field. Others said it was not dignified to make a controversial address in conjunction with an award ceremony. This experience pointed out the responsibility of scientists in continually testing the efficacy of any hypothesis, including toxicity testing models, is not necessarily widely applauded when strong financial stakes are involved.

THE SIGNIFICANCE OF THE APPEARANCE OF THE TERMS *ECOTOXICOLOGY* AND *ECOSYSTEM* HEALTH

I have stated[14] that I do not yet feel that I deserve to call myself an ecotoxicologist although I had been involved for two decades in multispecies and community-level testing. The reason that this remains an aspiration rather than an accomplished fact is that attributes recognized as fundamentally important to ecologists are not yet routine endpoints in the field of environmental toxicology. My assumption is that people label themselves ecotoxicologists to indicate that they reject the sole use of single species laboratory toxicity tests low in environmental realism as the primary means of estimating the effects of toxicants on ecosystems. A concomitant intent in the use of the term is almost certainly to relate toxicological testing more closely to ecosystem responses even though it is more of an aspiration than a reality at this time. If this is the case, then the term **ecotoxicologist** is used to indicate an expansion of the toxicity testing array now available without denigrating either single species tests or the events leading to the present evolutionary stage in the development of the field. Without doubt, there will be the usual lag of two or three decades before regulatory agencies embrace the new position. One example of this lag involves the use of various taxonomic groups in toxicity testing. Although it may be immodest to say so, I was using three different taxonomic groups (an alga, an invertebrate, and a fish) for aquatic toxicity tests carried out over 40 years ago, not only on a research basis but also in carrying out routine tests for industry, which then provided at least 50% of my extramural funding. This is the same administrative approach now adopted for effluent testing by the U.S. EPA.[15] I introduce this personal illustration as one example of a long delay by the U.S. EPA in accepting practices that were common in the scientific community many decades before.

CONCLUSIONS

The field of ecotoxicology appears to be moving conceptually along a different path from the regulatory development of the field, which is still focused primarily on single species tests, although tests at higher levels of biological organization are sometimes utilized. This dichotomy is evident in professional journals where some manuscripts fit well with current regulatory needs but others are so far removed from present practices that they might be talking about another planet. It is quite clear that most industries will respond primarily to regulatory needs and, thus, use the array of test methodologies currently favored by regulatory agencies. However, industry is becoming increasingly proactive and might find it cost effective to challenge the U.S. EPA's methodology by substituting a more scientifically justifiable methodology that is more appropriate to making judgments of the effect of toxicants on complex multivariate systems, especially at the landscape level.

Since the National Science Foundation funds very little research in ecotoxicology and the U.S. EPA's funding priorities are unlikely to rank research detrimental to regulatory policies at the top of its list, it is somewhat surprising that complex toxicity tests have flourished to the degree that they have. Some U.S. industries, especially those that are engaged in global marketing, are doing toxicity tests at higher levels of biological organization, not to meet regulatory needs in the U.S. but to meet challenges by green environmental groups elsewhere. I am proud that, in my own university, green engineering* is now well established. This was not driven by regulatory needs but, rather, by a perception of the way society's relationship to the environment has been evolving. In general, green engineering appears to be based on the assumption

---

* This was initiated by Dean Wayne Clough of the College of Engineering at Virginia Tech, who is now provost at the University of Washington, Seattle.

that good design removes stress from a system and that ecosystems are a necessary part of our overall life support system.

The trend toward toxicity testing at higher levels of biological organization is now well established. Despite a continued focus on single species toxicity testing, the U.S. EPA does have research going on at higher levels of organization and is sponsoring such research. Probably the most persuasive evidence is the ubiquity of multispecies toxicity testing, which, although not abundant anywhere, does have a fascinating global distribution. One need only consult the scientific literature to see the rate of growth for multispecies testing in the last decade.

The emphasis on ecosystem health, while more recently in vogue, also has a global distribution. It appears to be an irreversible trend despite the fact that the methodology and procedures for determining ecosystem condition and health are not yet as robust as multispecies toxicity tests.

Despite these trends, one still sees attempts to find an all-purpose indicator species for determining ecosystem condition and a most sensitive species for toxicity testing despite the improbability of estimates from a single species to a large complex system having even modest precision. A gulf between the research community and daily practitioners is inevitable. The former are interested in being at the cutting edge of the development of new knowledge, and the latter's interest is in the effective utilization of knowledge after it has been thoroughly tested and is in wide use and wide acceptance by the profession. There is an inevitable lag between the former and the latter, often as much as a decade or more since validation of any research investigators' results by others requires time. There is also the inevitable lag due to the process of publication, which, in many journals today, may take as much as 2 years. The question is, how much of a dichotomy will result in regulatory improvements and at what point will the dichotomy represent an insurmountable gulf between research and ultimate regulatory practices? Therefore, there should be a continuing effort to require validation of the accuracy of such predictions of indicator species or most sensitive species so that a large body of evidence exists on both points.

REFERENCES

1. Hart, W. B., P. Doudoroff, and J. Greenbank, *The Evaluation of the Toxicity of Industrial Wastes, Chemicals and Other Substances to Freshwater Fishes,* Philadelphia, PA: Waste Control Laboratory, Atlantic Refining (1945).
2. Kern, R. A., Arctic Haze Actually Industrial Smog? *Science,* 205:290 (1979).
3. Hirao, Y. and C. C. Patterson, Lead Aerosol Pollution in the High Sierra Overrides Natural Mechanisms which Exclude Lead from Food Chains, *Science,* 184:989 (1974).

4. Spencer, W. F. and M. M. Cliath, Movement of Pesticides from the Soil to the Atmosphere, in *Long-Range Transport of Pesticides*, D. A. Kurtz, Ed., (Boca Raton, FL: Lewis Publishers (1990), pp. 1–16.

5. Kurtz, D. A. and E. L. Atlas, Distribution of Hexachlorocyclohexanes in the Pacific Ocean, Air and Water, in *Long-Range Transport of Pesticides*, D. A. Kurtz, Ed., (Boca Raton, FL: Lewis Publishers (1990), pp. 143–160.

6. Johnston, C. A., N. E. Detenbeck, J. P. Bonde, and G. J. Niemi, Geographic Information Systems for Cumulative Impact Assessment, *Photogramm. Eng. Remote Sens.*, 54:1609–1615 (1988).

7. Hunsaker, C. T., R. L. Graham, G. W. Suter II, R. V. O'Neill, L. W. Barnthouse, and R. H. Gardner, Assessing Ecological Risk on a Regional Scale, *Environ. Manage.*, 14:325–332 (1990).

8. Suter, G. W., II, Endpoints for Regional Ecological Risk Assessment, *Environ. Manage.*, 14:9–23 (1990).

9. U.S. Environmental Protection Agency, Quality Criteria for Water, EPA 440/9–76–023, Springfield, VA: National Technical Information Service (1976).

10. Cairns, J., Jr., The Myth of the Most Sensitive Species, *BioScience*, 36(10):670–672 (1986).

11. O'Neill, R. V., D. L. DeAngelis, J. B. Waide, and T. F. H. Allen, *A Hierarchical Concept of Ecosystems*, Monographs in Population Biology 22, Princeton, NJ: Princeton University Press (1986).

12. Cairns, J., Jr., Will the Real Ecotoxicologist Please Stand Up?, *Environ. Toxicol. Chem.*, 8:843–844 (1989).

13. Cairns, J., Jr., The Threshold Problem in Ecotoxicology, *Ecotoxicology*, 1:3–16 (1992).

14. Cairns, J., Jr., Are Single Species Toxicity Tests Alone Adequate for Estimating Environmental Hazard?, *Hydrobiologia*, 100:47–57 (1983).

15. Wall, T. M. and R. W. Hamner, Biological Testing to Control Toxic Water Pollutants, *J. Water Pollut. Control Fed.*, 59:7–12 (1987).

CHAPTER 2

The Boundary between Ecology and Toxicology: A Sociologist's Perspective

Willem Halffman

Willem Halffman, of Science Dynamics at the University of Amsterdam in The Netherlands, visited my institution, as well as many others in North America, to determine why the fields of ecology and toxicology, which seem to a sociologist to be groups that would be working closely together on environmental pollution, were not nearly as interactive as an outsider might reasonably feel they should be. We discussed this book, and he decided to offer this chapter for publication. All other chapters were written by "insiders", but this one was written by an "outsider". This in itself makes it particularly interesting. For policymakers without a background in toxicology, it may well be the most interesting chapter. I have long held the opinion that young scholars should be given an opportunity to contribute their thoughts for a fresher perspective, even though they are inside the field. This idea of expanding the scope seems even more intriguing if the contributor is outside the field. Two toxicologists who reviewed this chapter both felt that the issues raised deserve attention, possibly because they furnish an illustration of how an outsider views the field. These reviewers suggested that the book begin with this chapter since it furnishes an interesting view on the evolution of the field and also the author's excellent handling of some contentious issues in a way that helps illuminate how science interacts with important social concerns. Both reviewers commented that Halffman's understanding of ecotoxicology is remarkable for a sociologist (or, for that matter, anyone outside the field), and his viewpoint is refreshing.

John Cairns, Jr.

INTRODUCTION

This chapter is not about different levels of biological organization, but about scientists who talk *about* different levels of biological organization. I am not an environmental scientist, but a sociologist who studies scientists. Like environmental sciences, my study is an interdisciplinary field consisting of historians of science, anthropologists, political scientists, philosophers, and sociologists, along with scientists who have taken some distance from their own discipline. This discussion of environmental sciences should be interesting for two reasons:

1. This view does not eliminate or marginalize the social and political aspects of day-to-day environmental science but makes them the focus; in this sense, an outsider may be able to identify mechanisms that those in the field might consider inconsequential.
2. This view contains naive reconstructions of what an outsider thinks environmental scientists do.

The impersonal tone of this chapter may be offensive to some since it discusses environmental scientists as if they were someone else. However, this tone is used so that I may remain neutral in the controversies that are described, even though some of these views may by now be outdated or even untrue. In the social science field, this is called the "principle of symmetry": the analyst does not choose sides in the (historical) reconstruction of scientific controversies. The evolution of scientific views, whether now considered true or false, must be considered in a similar conceptual scheme.[1] In fact, for this reason, some prominent scholars in my field have suggested that communication between scientists and social scientists be curtailed so that sociologists are not drawn into debates where their neutrality may be compromised because they become partisan in the controversies.[2-4] Unfortunately, in some cases, entire disciplines have refused to associate with sociologists because members have viewed sociological reconstructions as bad for their scientific status. In spite of these warnings, I have decided to write this chapter anyway, not because I want to become partisan but because my point of view as an outsider may help shape the debate.

ONE DECISION, SEVERAL SCIENCES

Governments often use science and its social legitimacy to guard against criticisms and counterclaims to their decisions. Scientific facts and expertise are used to set the limits of the attainable, the boundaries of nature. However, much to the dismay of governments and citizens, experts sometimes disagree. Experts may have personal stakes in the knowledge game, and the facts often seem less equivocal than expected. In addition, experts from different scientific disciplines may engage in debate or offer different perspectives on the problem or even take the liberty of reformulating the question.

One example of this situation is the practice of regulating the release of chemicals in the environment. Disciplines claiming to have the key to the environmental risk assessment of potentially toxic compounds range from chemistry, medical sciences, and microbiology to various subfields of toxicology and ecology. This discussion relates the recent history of chemical regulation in the context of toxicology and ecology and their mutual relations. Ecologists and toxicologists frequently disagree on pollution issues. However, some of them also make the promise of integrating the two fields, thereby creating ecotoxicology or toxicoecology. Whatever the name, a science was to be crafted that would bring together scientific facts on pollution and offer sound advice for chemical regulation on the basis of integrated knowledge — this (theoretically) would result in one field giving one decision on one environmental problem.

However, disciplinary boundaries limit the possibilities of integrative attempts as is evident in the reproduction of specific knowledge practices. [The term **knowledge practices** rather than **disciplines** or **specialties** is used to avoid discussions of what constitutes a discipline or whether particular fields are actually disciplines; this also maintains some neutrality since, for instance, Dutch (environmental) toxicologists present their knowledge practice as a full-blown discipline to defend their position in universities and research foundations.] These different knowledge practices construct different methods of dealing with the risks of pollution, in and with the policy practice, seeking support for different models, rationales, and tools for action. As a result, these knowledge practices affect decisions on what can be released into the air, soil, and water, and why different discharges can be released.

ECOLOGY, TOXICOLOGY, AND CHEMICAL RISK ASSESSMENT

This example is based on historical data that are still being collected on the situation that occurred when some laboratories focusing on aquatic biology tried to transcend the disciplinary boundaries of ecology and toxicology. The data compare the development of the boundary between the two disciplines from 1970 to 1990 in the U.S., The Netherlands, and the U.K., although the material is drawn mainly from the U.S. and The Netherlands and may not be considered by some as comparative at all since the data gathered are somewhat incomplete.

Ecology and toxicology are two different environmental sciences that both claim to be relevant to the problem of pollution, although the nature and strength of these claims vary both between the two fields and over time. Both fields have different structures. Toxicology, which largely originated from a pharmacological context, focuses on the effects of substances on humans. Several subfields have also developed, although the boundaries between these are not sharp and equivocal. For example, some toxicologists divide the field into the toxicology of drugs and drug combinations; dependence-producing substances; foodstuffs and food additives; pesticides; and industrial, environmental, incidental, accidental, forensics, military, and radiation toxicology. This division by area of application is further complicated by the introduction of division by approaches, such as acute or chronic toxicology.[5,6] Similar divisions in ecology are probably more widely known. Terrestrial vs. aquatic ecology, and population, community, and ecosystem ecology are two ways of structuring the field.[7,8]

Ecology is generally acknowledged as providing some of the basic metaphors and concepts for perceiving the environment; terms such as **ecologism**, **ecological problem**, and **ecosystem** have become part of everyday speech. This was the language the environmental movement used,[9,10] even after ecologists returned to their laboratories from their engagements in the practice and politics of environmentalism in the early 1970s. Ecologists had major difficulties in maintaining their central role and their high visibility among the environmental sciences. For example, in the U.S., the situation became especially pronounced with the passage of the National Environmental Protection Act (NEPA) and the subsequent establishment of the Environmental Protection Agency (U.S. EPA) in 1970. For example, the NEPA requirement for environmental impact statements pressed ecology to deliver practical results when ecology was not ready. At the same time, the Ecological Society of America tried to impose an ethics code (which failed) to add some legitimacy by listing certified ecologists.[7,11]

Since the failure in the 1960s of the International Biological Program (IBP) to centralize the control of ecosystems,[12] the major legitimizing strategy of most ecologists has been to refer to ecology as a "fundamental science", sometimes indicating that it is the science that would integrate environmental science, but, at any rate, referring to it on a basic level.[13,14] Even among other scientists, there seems to be a wide agreement on this status. This acceptance has assured ecology of a more or less stable legitimate niche in the science system and its funding channels, although the position was probably stronger in the late 1960s.

In 1969, the American Chemical Society (ACS) included the science of ecology in its comprehensive report on environmental problems. However, the chemists stressed that "a strong vein of chemistry runs throughout environmental chemistry and technology" and drew attention to analytic chemistry and measuring techniques. Three of the 72 recommendations in the report called for more ecological research — not merely implying this but explicitly using the terms **ecology** and **ecological** — and 7 called for biological research.[15,16] This report can be viewed as an attempt to secure a piece of the "war on pollution" for the ACS at a time when professional chemistry was confronted with stagnation and unemployment.[17,18] Environmental policy in the U.S. in the early 1970s seemed to be inspired by the idea that pollution could be eradicated. For example, with the 1972 amendments to the Water Pollution Control Act (Public Law 92.500), Congress defined a national goal of eliminating the discharge of pollutants into navigable water by 1985. The immense cost of the program was initially a bother to the administration and its austere budget. It took a few years and an upset Congress before the full implications and extravagance of this goal was realized.[19–22] However, the "war on..." seems to be a strong American policy metaphor that survived several similar disappointments.[23] In the early 1970s, most of the testing on environmental quality was based on chemical and physical parameters, and the stress by the ACS on analytical chemistry was strategically sound. The increasingly sophisticated techniques of analytic chemistry monitored the minuscule traces of pesticides and toxicants in general, and their transport and dispersal could be detected. In addition, analytic chemistry would prove to be of great use to the other group in ACS advisory committees: toxicologists.[6]

Toxicologists began building on their experiences with testing techniques developed for human toxicology, typically with rats as test animals, and their activities came to the forefront when environmental laws such as the U.S. Toxic Substances Control Act of 1976 were issued. Later, the new and growing field of environmental toxicology began a search for the most sensitive species to establish which substances would pose an environmental risk. By the end of the 1970s, the legitimacy of environmental toxicology was — and still is — mainly based on its use in the regulation of chemicals. The difficulty, then, became establishing toxicology as an academic discipline with a stable institutional structure with the right to perform fundamental research.[6] The establishment of toxicological expertise in regulatory practices has not been simple either, as "practical" toxicologists to some appear to be too closely tied to industry. A case in point is the accusation that two toxicologists violated governmental ethics in 1989. They had worked on a U.S. EPA advisory panel and accepted consulting jobs with industry after leaving the panel.[24] Especially in the U.S., this tie to industry is quickly perceived by environmentalists as "running with the devil". The expertise of toxicologists is quite often subject to counter expertise, as controversies over toxicant regulation are taken to court.[23,25,26]

However, this does not mean that ecology is not used in policy and regulatory practices. The acceptance of the field of ecology generally depends on the acceptance of the idea that the environment is a strongly intertwined whole.[6] Although this idea can be interpreted in various ways, policymakers and toxicologists generally agree with ecologists that the ultimate end of all efforts is "ecological", i.e., to protect the whole of the environment. Since many ecologists feel secure that the field is being accepted, they have again turned their attentions to practical environmental questions, where they were confronted with the well established position of the more reductionist toxicologist. The differences between ecologists and toxicologists have continued since ecologists have failed to develop useful tools for regulatory purposes and toxicologists continue to base many of their ideas on human toxicology. These differences are illustrated by this quote from a U.S. EPA official in 1981:[27]

> Ecologists often fall short when it comes to applying the theory and findings of their relatively young science in useful practice to meet society's needs for assessment of the environmental impacts of toxic pollutants. Environmental toxicologists are increasingly having difficulty in trying to convince society's decision-makers what the results of their test methodologies in simple systems really mean in a complex, highly interactive ecological world.

Beginning in 1969, the call for "interdisciplinary approaches" and "combined efforts" began.[15] This integrative science combining toxicology and ecology seemed a noble goal and the right way to allow these sciences to address the pollution issue.[27] However, toxicology and ecology did not merge for many more years.

THE BOUNDARY

The fields of ecology and toxicology have discontinuities in organizational structures such as laboratories and funding organizations, in the structure of their respective journals, and in the structure of research programs and theories. These discontinuities suggest continued, separate use of the terms **toxicology** and **ecology**, which are in themselves already a way of establishing difference.

(Environmental) toxicology and ecology have distinct research programs that identify different problems and that take different things for granted. Toxicologists assess what happens to an organism when it is exposed to a chemically defined substance or mixture of substances. (A traditional definition is: "...the study of injurious effects of substances on living organisms. It is a subdivision of pharmacology. Pharmacology is the study of molecular, chemical interactions between substances and biological objects."[5]) The organism sets the limits of the investigation. Even though a group of organisms of the same species might be investigated, the main purpose is to establish the response of a statistically defined standard organism. Initially, toxicology dealt with the entire organism, and typical research questions were: When will 50% of the test animals die? How long does this take? What concentration or dose does it take? In environmental toxicology, the investigations shifted towards physiological[24] and even biochemical modes of analysis, which were considered to be more fundamental levels of explanation. Specific effects of substances on the nerve system were traced in "neurotoxicology", which began with research in entomology on the study of insecticides that blocked insect neurotransmitters. More recently, toxicologists have turned to "immuno-toxicology", that deals with deleterious effects on the immune system.[6]

Ecological studies do not typically deal with the organism itself but stress the relation of organisms to their environments. The level of aggregation is higher when treating populations or coexistent groups of populations. There also is an "ecophysiology", but, here too, attention is focused on (physiological) interactions with the environment. In addition, ecologists who follow specific substances and their ecological fate are generally preoccupied with naturally occurring substances, such as nutrients (elements such as carbon, nitrogen, or phosphorous), that are considered as the fundamentals of the ecosystem. Data gathered on the "natural state" are often used as the base line for assessing effects of anthropogenic interference.[7]

In addition to these different demarcations of what is problematic, toxicologists and ecologists use different tools, methods, theories, and concepts. For instance, toxicologists isolate an organism in an experimental setting to control and, thus, neutralize ambient conditions, whereas these are precisely the elements in which ecologists would be interested. Even when an environmental toxicologist attempts to imitate natural conditions, the method is still the assessment of toxic effects on individual organisms under controlled conditions.[28]

In addition to the differences in tools, methods, theories, and concepts, scientists in each of the two disciplines view them as two different fields. Statements that indicate differences and the nonintegration of ecology and toxicology can be found throughout the literature,[29,30] and scientists generally maintain a rigid classification of the ecologists and toxicologists whom they know. These boundaries consist of sharp group definitions, indicating who is "in" and who is "out", though the divisions within the group may be less rigid but more complex. Moving into other areas, across boundaries recognized by ecologists themselves, is perceived as difficult and may lead to self-exclusion from other fields.[8] The group definitions might not indicate who precisely is the relevant other (from whom to seek support or acknowledgment), but they do indicate that toxicologists and ecologists are generally not relevant others.

Another difference can also be found in the journals that each group of scientists considers relevant. For instance, a toxicologist could exist without reading *Ecology*, but would have to confer with *Environmental Science and Technology* to stay in touch with developments (personal observation). The citation structure of journals in (aquatic) environmental science (revealed through citation mapping techniques) indicates that toxicology and ecology are quite distinct and have been since the mid-1970s. Only towards the end of 1980 did the links with toxicology and hydrobiology become more pronounced — aquatic ecology being well integrated with hydrobiology in terms of this citation structure.[31]

A fourth example of the discontinuity between the two fields is evident in the Dutch Biological Research Foundation, BION,[32] which regulates and secures funding for biology. BION also attempts to structure and coordinate Dutch biology through a network of peer groups that are organized hierarchically into sections, divisions ["werkgemeenschappen" (terms here are the same as those used by BION)], and research groups ("werkgroepen", headed by a senior scientist and dealing with laboratory work). Discussion groups initially dealt with plant pathology (1983), then soil biology (1984, with possible contacts to further toxicological research, and later animal experiments (1989) (BION Annual Reports). This scheme has undergone some changes since the creation of BION in 1970, but it has maintained its hierarchical, tree-like structure with only a few crossovers.[33,34] When BION was criticized for this tree-like structure, alternative structures such as the discussion groups and specific research programs in the organization were emphasized. However, the basic functional element in BION is the division, which reviews proposals and allocates the scarce funds to projects. In this design, prominent members in the field of Dutch biology have to reach some form of agreement on what constitutes a good proposal, what the research priorities are, and (not entirely unimportant from the perspective of this discussion) a definition of the field. This definition is more or less explicit in the research programs that are written and is also evident in the choices of members accepted into the divisions and in the proposals that are accepted as falling under the jurisdiction of a division.

Ecology groups joined BION early, most in 1971. Toxicology did not become an independent division until January of 1990, when it became part of the section of animal physiology. This created the widest organizational gap possible in BION between ecology and toxicology.[35]

Since 1982, a steering committee ("Dwarsverbandcommissie", literally: cross-linkage commission) has existed between BION and FUNGO, the parallel funding organization for medical research. This committee has coordinated toxicological research and managed several priority status funding programs for toxicology such as BION priority in 1983 (following the advice of the cross-linkage commission) and the special ZWO program in 1987 (ZWO: fund for pure scientific research, later NWO).[35] BION has had its own project group on toxicology, which was in close contact with the steering committee and constituted BION's part of it. This project group was the direct predecessor of the toxicology work community and functioned since its inception in animal physiology, but with close personal links to the study group of metabolic physiology.

The structure of BION facilitates relations and contacts between scientists, although with rather variable success. For instance, some Dutch toxicologists find the BION network important for contacts, while others depend on alternative channels (such as the new national graduate training program in toxicology) and consider BION merely a way to organize extra funding. More important, however, is that the criteria for funding demand standards of good research and research priorities and are defined separately for toxicology and ecology. However, research groups find it difficult to integrate both approaches since it requires adapting research not to one but to two sets of standards if the groups apply for funding to two different divisions.

In addition to these difficulties is the fact that toxicology and ecology within Dutch universities are widely separated physically. Many toxicology laboratories are not even part of the biology departments. This may not necessarily prevent contacts between the two knowledge practices and has the advantage of reducing competition for the limited resources of the department; however, it does not increase the likelihood of contact or cooperation either. For example, the study of toxicology is included in veterinary sciences at the University of Wageningen and at the University of Utrecht (although some have double appointments in biology) and in the chemistry department at the University of Amsterdam. However, some research institutes of the government do not appear to follow the disciplinary partitioning of universities. This situation does not appear to be as pronounced in the U.S. as in The Netherlands.

These boundaries and partitioning are not absolute and are understood only as they relate to each other. Toxicology and ecology are university-based subjects, generally part of the environmental sciences. When they are organized (as with the example of BION), they are sometimes part of the same larger organizational structures. Even though the boundaries with other subjects, such as sociology, may be stronger and sharper by comparison, differences between ecology and toxicology are still substantial.

However, can these differences be considered "boundaries"? Differences can be listed that may lead to specific patterns of inclusion and exclusion. Understanding these inclusions and exclusions requires a closer view of specific historical differences. Specific examples point out the way boundaries are set, maintained, and made routine. Scientists have attempted to move from one field to another or to link elements from both fields. Such efforts began in Europe in the early 1970s.

THE PROGRAM OF ECOTOXICOLOGY: THE BIRTH OF A CONCEPT

In June 1969, René Truhaut, a French toxicologist with a pharmacology background, coined the term **ecotoxicology** — "a new branch of toxicology" studying toxic effects "to the constituents of ecosystems". He did so in the context of an ad hoc committee of the International Council of Scientific Unions (ICSU), and the term was adopted by the Scientific Committee on Problems of the Environment (SCOPE) that was subsequently formed in the ICSU. Truhaut chaired a working group on ecotoxicology that was charged with drafting a precise definition of the new field.[36] However, the formal definition that was commonly adopted in the following years was devised by the new SCOPE preparatory commission for the Project on Ecotoxicology in workshops in Brussels and Neuherberg, Germany, in 1976. The report, "SCOPE 12", was issued in 1978 and contained the following definition:

> Ecotoxicology is concerned with the toxic effects of chemical and physical agents on living organisms, especially on populations and communities within defined ecosystems; it includes the transfer pathways of those agents and their interactions with the environment.[37]

The definition was designed to avoid overlap with the study programs of the World Health Organization and to differentiate the "new" or "emerging science" from classical toxicology, which has its focus on the effects on individual organisms and human health.[36,37] The debate still continues on whether ecotoxicology has been successful with this differentiation.

However, a new program could hardly be launched in a better context — SCOPE and the ICSU provided a worldwide network of scientific statespeople whose activities were funded by eminent sources.[37] The term **ecotoxicology** quickly became more than a quote in a report. In 1977, even before the SCOPE report was issued, the journal *Environmental Quality and Safety* reappeared as *Ecotoxicology and Environmental Safety* "under the Auspices of the International Academy of Environmental Safety" (IAES). The journal had close relationships with the Society of Ecotoxicology and Environmental Safety (SECOTOX) and became its official journal in 1982. (The Society only began using the word **ecotoxicology** in 1983.)

Yet another example of the fruitful ground on which the concept of **ecotoxicology** fell was Jan G. P. M. Smeets. Trained in forestry at the Agricultural University of Wageningen, The Netherlands, he was president of IAES, involved with SECOTOX from its beginning, and responsible for pollution prevention at the European Community (EC) Commission Administration. In this position, he was involved in the preparation and implementation of the Community Directive on control of new chemicals[38] and wrote several articles for *Ecotoxicology and Environmental Safety* on EC environmental policy.[39,40] The concept of ecotoxicology quickly spread throughout environmental science and is now regularly used. In fact, the term appeared in publications around 1980 to 1982 in the U.S.

However, what was "ecotoxicology" to investigate? The "SCOPE 12" definition and the program formulated in the report were based on the idea that toxic substances were to be evaluated at their point of action, the environmental receptors. In order to do so, four processes had to be analyzed:

1. release of toxicants in the environment — their quantities, forms, and sites
2. transport and chemical transformation of these substances throughout the environment
3. exposure of the receptors
4. response of these receptors[41]

During its development, the research program that had been associated with ecotoxicology in its original formulation was soon forgotten. The concept took on numerous meanings as it became more popular; however, they can be largely placed into two interpretations. An inclusive meaning used the term for everything related to the effects of toxicants in the environment, including environmental toxicology and its research into effects on individual organisms. Nothing in the SCOPE definition excludes this interpretation. However, the definition does state that ecotoxicology focuses especially on higher levels of aggregation: "populations and communities within defined ecosystems". This more exclusive interpretation has been stressed mainly by ecologists who increasingly used the term in the sense of the integrated field of ecology and toxicology — the study of toxic effects on these higher levels of integration, with special attention to the interaction between species in transforming and accumulating substances.[42]

Ecotoxicologists who adopt the exclusive interpretation of the term have attempted to establish the more narrow definition and keep the term "clean", especially for toxicologists who use this "buzz word" for their single species research. In some instances, these attempts were explicit — in 1989, the journal *Environmental Toxicology and Chemistry* published an editorial featuring a range of possible interpretations and defining "the real ecotoxicologist" — "one who assesses the effects of toxic substances on ecosystems".[43] Ecotoxicologists continue to establish and defend the boundaries of their own research programs, even if they are weak and violated regularly. Even so, ecotoxicology has spread, however vague the definition. Dutch government reports and funding agencies,[13,32,33] for instance, sometimes speak ecotoxicological language. Since 1983, a handbook on ecotoxicology has been available in The Netherlands for training students,[30] and, throughout the world, several laboratories claim to practice ecotoxicology and use the term in their names (although this seems to be the case more in Europe than in the U.S.).

However, the old boundaries of ecology and toxicology still remain ingrained in ecotoxicological programs; integrating research across these boundaries has proven difficult. Ecotoxicology still has not been able to establish itself (at least not to the extent that toxicology and ecology have), and supporters of the program stress that "real" ecotoxicological research is still scarce.[30,43] In addition, ecotoxicological research is often identified as such after the fact (for example, accumulation of DDT in food webs and the methylization of mercury in the Minimata Bay disaster in the 1970s).

Most ecotoxicological research to present has used previously existing methods that were expanded into this new field of research, rather than developing new or integrated practices. Research into the transport and chemical transformation of substances has become the field of environmental chemistry, which makes use of analytical techniques; environmental and chemical technology focuses on releases into the environment; and toxicology is the dominant factor in the remaining types of research. Since 1977, chemistry has become increasingly important in environmental toxicological research. Techniques have been introduced for use in predicting the biological activity of chemicals from their chemical structure or parameters describing this structure, respectively called structure-activity relations (SAR) and quantitative structure-activity relations (QSAR). SAR and QSAR have been used successfully in pharmacology, sometimes with spectacular results.[44–46] Some of this research resembles ecotoxicological work, as claims have been made concerning predicting ecosystem parameters,[47] and a standard handbook of ecotoxicology now includes a section on QSAR.[42] The technique has been welcomed by industry for use in regulatory testing because of its low costs[48] and has been accepted and developed by U.S. EPA.[23,47]

TOXICOLOGISTS, ECOLOGISTS, AND REGULATORY PRACTICE

TESTING FOR DECISIONMAKING

If regulatory agencies such as the U.S. EPA accept a method in its official set of toxicity testing techniques, this does not *directly* mean more funds for associated research. Most testing research is part of the product development cost to industry and is generally not public knowledge. In addition, such research is routine and restricted by its specific purpose; it is anything but innovative. Even though this type of research provides jobs in industrial and specialized testing labs, it is not, in itself, a major channel for pioneering academic research. In addition, in a field where environmentalists, governmental agencies, and industries mobilize expertise and counter expertise for mutual benefit in high stake conflicts, an expert who engages in such regulatory research might have his or her credibility questioned.[23–25]

Government agencies have increasingly given their testing procedures various systems of standardization to strengthen regulatory decisions. This results in a limited range of accepted tests for specific environmental parameters. Detailed testing protocols are defined and good laboratory practices (GLP) have also been issued.[6] These standards are mainly dependent on government organizations, as indicated by the fact that they are substantially different within national and international organizations. Laboratories that are oriented to the international market must deal with the different regulations of such organizations and nations as the EC, OECD, Japan, U.S. EPA, and U.S. FDA (which designed the first GLP rules in 1978; the OECD version followed in 1982).[8] With the support and alliance of legal procedures, scientists in government, and leading scientists in the field, these procedures have been made into obligatory passage points[49] for testing of chemicals. (The term **obligatory passage point** was introduced by Bruno Latour and is used to indicate a situation where one must pass through a particular system to achieve a certain goal. For instance, Checkpoint Charley was used as an obligatory passage point when one wished to move from West to East Berlin.) This system requires the development of these procedures, the comparison of results between laboratories, the reiteration of research results until they speak the same language, and the legitimation of these results by reduction to what are considered to be more fundamental levels of explanation. In short, this requires a great deal of research that can at least partly be oriented to the frontiers of the field. In addition, actual research offers a new perspective to the social legitimacy of "doing something about pollution", which is important in motivating and involving environmental scientists.

ECOLOGISTS FOR DECISIONMAKING

In the first half of the 1970s, environmental regulatory testing focused on the control and monitoring of wastewater effluents and ambient water quality standards; the testing was often performed by employees of wastewater treatment plants. Ecologists attempted to convince these engineers that ecological parameters were superior to the analytic methods used earlier. Some ecologists went to great lengths to develop techniques that would seduce nonecologists into monitoring such things as species diversity indexes. In one case, John Cairns, Jr., an ecologist and Director of the University Center for Environmental and Hazardous Materials Studies at Virginia Polytechnic Institute and State University in Blacksburg, Virginia, explicitly summed up typical engineering objections, such as the engineers' lack of biological knowledge or the lack of numerical data from biology, and tried to refute these by presenting a ready-made measuring technique. The instructions included detailed instructions for sampling and technical information on dredges, mathematical formulas, and even a full Fortran computer program to calculate one simple species diversity index that would allow clear and quantitative evaluation of water quality.[50] Diversity was viewed as an indicator of ecosystem stability, as was common for ecologists influenced by E. P. Odum in those days.[12]

Before the U.S. EPA took more control of water pollution programs (after the 1972 amendments to the Water Pollution Control Act), the only way to introduce these practices to regulatory agencies was to convince the engineers in the Water Pollution Control Federation (WPCF). The WPCF published *Standard Methods for the Examination of Water and Wastewater* in cooperation with the American Public Health Association and the American Water Works Association. In 1971, the extensive list of testing methods included a small section on "Biological Sampling and Analysis" (Part 600), but none of these criteria concerned aggregation levels higher than individual species.[51] After 1972, when the U.S. EPA increasingly centralized the testing practice and its requirements, the U.S. EPA became the main focus of attention for ecologists. As noted before, the U.S. EPA's criteria were initially defined entirely on the basis of physical and chemical water parameters based mainly on toxicological research. For example, the U.S. EPA water quality criteria, issued after the Water Pollution Control Act amendments of 1972 and applicable to effluents, included the parameters of temperature; disinfection; concentrations of phosphate, mercury and other heavy metals, nitrate, and pesticides; and measurements of dissolved oxygen, radioactivity, etc.[52] This list was even more limited to chemicophysical parameters than the wide range of tests recommended in the WPCF list. Testing continued in this manner in the U.S. until biomonitoring techniques based on specific "indicator species" were supported and standardized by the U.S. EPA in 1987. American biologists tried to convince the U.S. EPA with the argument that these techniques were used successfully in Continental Europe and the Soviet Union.[53-55]

However, this was not what ecologists wanted. To them, single indicator species were incomplete and inadequate indicators of water quality, and they wanted more ecological parameters to be included.[56]

ECOLOGISTS FOR REALISM

Ecologists also protested the U.S. EPA testing practices developed under the U.S. Toxic Substances Control Act of 1976. The regulatory system that was developed under this law limited the toxicity tests to the effects on single species and laboratory bioassays.

Ecologists opposed to the sole use of single species testing used the following argument:

> ...such tests are relatively meaningless since such pollutants as pesticides are applied, in practice, to the community and ecosystem levels of organization and not to the organism or population level.[57]

This argument is similar to the one used by ecologists in relation to ecotoxicology: individual organisms are but parts of a whole that is to be protected; various interactions and chemical transformations between species may alter the effects of pollution entirely. Therefore, ecologists judged single species testing to be too crude a model and wanted a more complete model that would represent an ecosystem instead of only one species of it, thereby guaranteeing a more accurate prediction of the environmental effects of chemicals. This would not necessarily exclude an increasing number of chemicals from use even though the adjusted results might sometimes be more strict, sometimes more lenient. It was claimed such a model would streamline the regulations, eliminating "false positives" (rejected chemicals that are actually safe) and "false negatives" (accepted chemicals that are actually dangerous).[58]

By the end of the 1970s, mathematical models had lost much of their credibility, although many ecologists continued to propose them. However, an alternative model came into use: the microcosm — "... reduced scale models of natural ecosystems or portions of natural systems housed in artificial con-

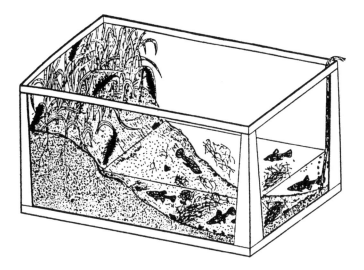

Figure 1 A 1971 microcosm (then called "model ecosystem") from Metcalff and colleagues. (From Metcalff, R. L., Sangha, G. K., and Kapoor, I. P., *Environ. Sci. Technol.* 5(8):709–713, 1971, with permission.)

tainers kept in a controlled or semi-controlled environment."[59] The term **microcosm** appeared in a publication as early as 1963 concerning a study of microbial metabolism in sewage oxidation ponds. **Microcosm** has been used interchangeably with **microecosystem,** a term that is still frequently used, for instance, by Dutch ecologists.[60] According to Limoges and colleagues,[61] the term was used as early as 1887[62] and reappeared as an experimental notion, rather than an exact term, in a 1957 article by Odum and Hoskins.[63] However, the concept is even older than this as used by Paracelsus (1493–1541): "Man is a small replica of the great world about him, and within him are represented all parts of the universe...."[64] The parallel is remarkable.

The microcosm design uses several selected organisms rather than a single one and interactions are used as ecosystem parameters. This mimics the environment more closely as the relevant parameters are ecosystem parameters, not just individual species responses.

Early microcosms were often ecologically complex aquariums with plants, fish, natural sludge, and microorganisms. Figure 1[65] shows a microcosm that was proposed for use in testing under the U.S. Department of Agriculture regulations governing pesticides. Microcosms were soon simplified and used, for instance, a few species of microorganisms that represented a few trophic levels. Taub[66] proposed systems that contained one species of algae, one protozoan, two bacteria, and a nitrate nutrient solution. Figure 2[67] contains three trophic levels: primary producers (algae), primary consumers (*Daphnia magna*), and reducers (bacteria). These microcosms were supposed to become the "white rats" of ecology.[68,69]

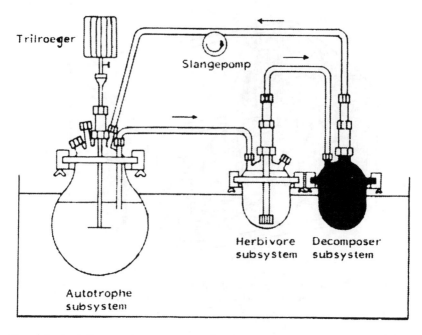

Figure 2   A 1985 microbial microcosm (then called "microecosystem") from Kersting. (From Kersting, K., *Vakblad voor Biologen*, 6(13/14):118–119, 1985, with permission.)

It is important to note that in both cases, biomonitoring and toxic substance testing, the strategy of ecologists was similar: the point was not to remove single species testing. Its value was repeatedly acknowledged and Cairns even rebuked some of his allies for scaring U.S. EPA officials too much when debunking single species tests overenthusiastically. The prime goal of ecologists was not to move single species testing out, but to move microcosm testing into regulatory practice.[59,70,71]

Apart from defending the use of microcosms themselves, ecologists tried to have influential sources speak the same language. They tried to interpret environmental laws in their support — the Clean Water Act of 1977 was interpreted as mandating "a more holistic approach"[70,72] — and a committee of the National Research Council produced a report defending the use of microcosms in 1981.[73]

Eventually, the Office of Toxic Substances of the U.S. EPA asked Oak Ridge National Laboratory to write a report evaluating microcosm use. The report was published in 1981 also and contained substantial criticism of U.S. EPA testing regulations. However, the report also concluded that only a few "multispecies toxicity tests" could be made available in the near future, most needing substantially more research. Not one was considered fit at that time for immediate use.[74] The report also contained a small section evaluating the potential of mathematical models, with little positive result. Ecologists were in a Catch-22: for microcosms to be accepted by regulatory authorities, they needed standardization; to obtain standardization, microcosms seemed to need acceptance by regulatory authorities.

Before 1983, toxicologists and the U.S. EPA paid little attention to criticism of ecologists. In that year, Cairns enlisted the help of the Ecological Society of America and the Society of Environmental Chemistry and Toxicology (SETAC) in funding a workshop on multispecies testing at Virginia Polytechnic Institute and State University. The entire scenario was to be discussed. Convincing industry would not have been enough in itself if the U.S. EPA did not follow. The U.S. EPA had to be convinced of the toxicologic data, and U.S. EPA officials had to know and accept the techniques used. However, finding an ally in industry might pressure the U.S. EPA. SETAC, which published the papers in 1985 as the first volume of its special publications series, provided a perfect forum since its membership overlapped industry, science, and government agencies. One of the proclaimed goals of SETAC was to provide possibilities for contacts and rapprochement between these parties, not unlike the Water Pollution Control Federation mentioned earlier.[75] As for industry, only Monsanto had already been experimenting since 1979 with a microcosm pond.[74] The volume that reports on this workshop shows that researchers and representatives from both the U.S. EPA and industry participated. Several objections to multispecies testing surfaced, besides lack of standardization. One of them was the higher cost of multispecies testing. Calculations were given to show that one well-chosen microcosm test design would cost about as much as the series of minor single species tests it would replace. The main element in the cost was the expertise of laboratory personnel, which was required to be higher in multispecies testing and might be related to nonstandardization.[76] Arguments went both ways: What was really measured? Where was the validation? What was to be measured? What is the "normal state"? The 1981 report[74] to the U.S. EPA had used a list to evaluate multispecies tests: cost per test, documentation, generality, rapidity, realism, (rejection) standards, replication, reproducibility, sensitivity, social relevance, standardization, statistical basis, training-expertise requirements, and validity. These are precisely the kind of arguments used by ecologists against single species testing. Thus, the "white rat" status of microcosms was clearly not accepted, nor was the logic behind it condoned:

> ...we may be able to distinguish certain universal ecosystem properties measurable in all systems and, by studying these properties and their response to toxic chemicals, make inferences that would be meaningful in any system. ...Such model ecosystems, consisting of a few species of bacteria, algae and invertebrates, have no natural counterparts; in a strict structural sense, they are totally unrealistic, and yet they exhibit features such as succession, metabolic balance, and homeostasis that are characteristic of all terrestrial and aquatic ecosystems.[69]

If one does not accept the reality of metabolic parameters of ecosystems on a par with metabolic parameters of organisms, then this would be a sufficient argument against microcosms and for single species tests. "Unfortunately, the universal ecosystem properties of which we are currently aware are of little recognized social or economic relevance."[69]

Dying animals are clear, measurable, and highly visible end points, unlike gross primary production of an ecosystem. Little social support could be expected for these parameters that only ecologists really understood. To make things worse, ecologists did not even agree on which of the many parameters available was to be used in multispecies tests. Nevertheless, even the most abstract of environmental problems, such as the greenhouse effect and the destruction of the stratospheric ozone, could gain support. However, in this case, major political decisions were made on the basis of (controversial) evidence of a possible hole in the ozone layer over the Western Hemisphere and the warning of an increased risk of skin cancer.

In addition, this problem has to be viewed against the larger background of U.S. EPA policy. When, in 1976, Douglas M. Costle became U.S. EPA Administrator under the Carter Administration, he sensed a decreasing support for environmental issues. This, combined with the threat of government re-form, was felt by Costle to endanger the future of the U.S. EPA. In response, Costle refocused the attention of the U.S. EPA to the protection of public health that "replaced the maintenance of ecological balance as the EPA's central objective."[23] The U.S. EPA jumped on the cancer bandwagon and gave it a nice push itself. Only after early 1985, when Lee M. Thomas became the U.S. EPA's administrator, the priority returned to ecological issues and became concerned with global issues such as ozone depletion and the greenhouse effect.[23,77] Thomas established the Ecotoxicity Subcommittee of the Risk As-sessment Council to identify appropriate ecological endpoints and develop ecological risk assessment guidelines.

As Jasanoff noted, it would be distorting the picture to present the U.S. EPA as a monolith with one will and intention.[78] The U.S. EPA was not antiecology, as it supported ecological research, even on microcosms, for instance, at its Environmental Research Laboratory in Athens, Georgia,[69] and through research grants.[79] In other words, ecological testing methods may have been included or excluded in the different offices of the U.S. EPA, with some offices being less ecology minded than others. Remarkable differences may also have occurred on an individual level. That is, both the introduction of microcosms into the Office of Toxic Substances and the experiments with mesocosms in the Office of Pesticide Programs seem to have been almost personal projects of relatively small groups of individual officials, which is not exceptional in bureaucracies. For example, when the national ambient air standards for ozone were reformulated in 1979, the ad hoc advisory boards were dominated by medical scientists so that the discussion was dominated by concern for potential human health effects.[23] McIntosh comments on a similar situation when he notes of the Presidential Council of Environmental Quality that "its major figures have not been ecologists..."[7]

EPILOGUE

In 1988, single species testing was still dominant in spite of all the scientific arguments ecologists had offered. Single species testing was too embedded in regulatory practice to be removed easily; it was inexpensive, was formally standardized, was based on a long tradition of mammalian tests designed to protect humans, had easily quantifiable endpoints, could be used by laboratory personnel without ecological knowledge, and represented substantial economic and regulatory forces. It offered relatively clear, straightforward results and a large number of firms who specialized in testing had major economic interests. However, multispecies testing was still defended as offering more "environmental realism". Researchers were hoping for support from the closer association of ecologists and environmental toxicologists in the development of a true ecotoxicology.[80]

ONE SCIENCE, SEVERAL DECISIONS?

Ecology and toxicology have several boundaries coinciding, leading to "boundary alignment". The pattern of exclusion in testing practices coincides with patterns in research programs and with several differences on institutional levels, such as funding organizations.

Both toxicology and ecology are well established and have integrated institutional mechanisms that will assure their continued existence — called "institutional alignment".[81] Both knowledge practices dictate journal outlets, research facilities, channels of funding, relations with government, and other audiences.

While the institutional alignment assures and integrates various resources for a group of researchers, the boundary alignment sustains this process for a specific group, with particular patterns of inclusion and exclusion. As the boundaries of knowledge practices are redrawn, they define the scope of acceptable questions in a discipline, how they are to be formulated, how problems can be solved, and who is to solve them.

This does not mean that the differences among knowledge practices (or the "stability" of the difference as some might want to call this) can be narrowed to one, single, specific boundary. It is not the different research programs that explain why ecology and toxicology do not integrate, for attempts have been made to design integrated research programs. It is not because of the separate journal structure, for here, too, crossings have been attempted. The difference between these two knowledge practices is maintained through the interaction and mutual reinforcement of these different boundaries.

On the other hand, this cannot be explained by merely stating that "everything plays a role", but when different patterns of inclusion and exclusion of a discipline coincide, they reinforce those differences. Thus, different knowledge practices remain distinct and continue their development separately, maintaining different philosophies of what the(ir) world is like.

However, the different types of boundaries carry varying strengths. Some may be stronger than others and may be more powerful resources for researchers to maintain control over their domain and who is part of it. For instance, boundaries become reinforced and knowledge practices more isolated when governmental agencies require legitimacy in decisionmaking. Expert knowledge must appear depoliticized to be of any use in the policy process. An alliance with the government in the testing practice may then become powerful in maintaining the boundary between toxicology and ecology.

The framework of boundary alignment/institutional alignment cannot be reduced to a few crucial factors nor explained casually. The system is only a set of interpretative categories, ordering elements like the standard testing of chemicals along two analytical axes. One axis interprets this testing practice as a crucial element in the alliances of toxicologists; the other, the boundary aspect of the testing practice, draws attention to how this resource is tied to an exclusive group of researchers.

All this does not imply a universal conspiracy of toxicologists or ecologists. The boundaries can be seen in mundane day-to-day activities such as the approval of a proposal or a journal article but can also be seen in the more planned efforts to assure exclusive professional domains, as in the lists of certified experts that American ecologists compose.

One final provocative statement. The preferences of a discipline may not be a purely negative implication of boundaries. Even though boundaries limit the scope of a discipline, they also proscribe a necessarily limited set of strategies, techniques, and models. A test system for chemicals that would question all aspects simultaneously would never allow a decision on the use of a chemical substance, if such a system revealed anything understandable at all. In addition, a unified, boundless, and "nondisciplinary" science would imply one unified reality, one objective truth. Decisions could never be made under such a totality. The issue is not "bound or boundless" but "which boundaries and where".

ACKNOWLEDGMENTS

I want to thank my colleagues at Science Dynamics for their comments and stimulating discussions, especially Rob Hagendijk, Kaat Schulte, Chunglin Kwa, and the discussants at the Science Dynamics Internal Progress Conference, May 1992 and at the 4S/EASST Conference in Gotheburg, Sweden, August 1992. I also wish to thank Steven Shapin and Tom Gieryn for their remarks on the research proposal, Steven Yearly, the toxicologists and ecologists who took time to answer my fundamental questions, and to the extremely helpful people of NWO and BION, such as Marga Verschoor who supplied the BION terms used in this chapter. I also want to thank John Cairns for offering me the opportunity "to talk back" and for the adaptation of the text.

REFERENCES

1. Bloor, D., *Knowledge and Social Imagery,* London: Routledge & Kegan Paul (1976).
2. Scott, P., E. Richards, and B. Martin, Captives of Controversy: The Myth of the Neutral Researcher in Contemporary Scientific Controversies, *Sci. Tech. Human Values* 15(4):474–494 (1990).
3. Collins, H. M., Captives and Victims: Comment on Scott, Richards, and Martin, *Sci. Tech. Human Values* 16(2):249–251 (1991).
4. Scott, P., E. Richards, and B. Martin, "Who's a Captive? Who's a Victim? Response to Collins's Method Talk, *Sci. Tech. Human Values* 16(2):252–255 (1991).
5. Ariëns, E. J., A. M. Simonis, and J. Offermeier, *Introduction to General Toxicology* (revised printing), New York: Academic Press (1976).
6. Groenewegen, P., *Scientists, Audiences and Resources*, Science Dynamics Dissertation, Groningen: Van Genderen (1988).
7. McIntosh, R. P., *The Background of Ecology: Concept and Theory,* Cambridge, Massachusetts: Cambridge University Press (1985).
8. Hagendijk, R. and J. Cramer, Intellectual Traditions as Cognitive Constraints, *Soc. Sci. Inform.* 25(3):703–723 (1986).
9. Bramwell, A., *Ecology in the 20th Century: A History,* New Haven, Connecticut: Yale University Press (1989).
10. Pepper, D. *The Roots of Modern Environmentalism,* London: Croom Helm (1984).
11. Nelkin, D., Scientists and Professional Responsibility: The Experience of American Ecologists, *Social Stud. Sci.* 7:75–95 (1977).
12. Kwa, C., *Mimicking Nature: The Development of Systems Ecology in the United States, 1950–1975*, Science Dynamics Dissertation, The Netherlands: University of Amsterdam (1989).
13. Aart, P. J. M. van der, J. van Andel, and J. W. Woldendorp, *Programma voor Oecologisch Onderzoek [Program for Ecological Research],* The Netherlands: BION (1988).
14. Odum, E. P., The Emergence of Ecology as a New Integrative Discipline, *Science* 195(4284):1289–1293 (1977).
15. American Chemical Society (ACS), Committee on Chemistry and Public Affairs, Committee on Environmental Improvement, *Cleaning Our Environment: The Chemical Basis for Action,* Washington, D.C.: ACS (1969).
16. American Chemical Society (ACS), *A Supplement to Cleaning Our Environment: The Chemical Basis for Action, Priority Recommendations with Supporting Discussion,* Washington, D.C.: ACS (1971).

17. Gillette, R., ACS: Disgruntled Chemists Seek New Activism from an Old Society, *Science* 173:1218–1220 (1971).
18. Thackray, A., J. L. Sturchio, T. P. Carroll, and R. Bud, *Chemistry in America, 1876–1976: Historical Indicators,* Dordrecht: D. Reidel (1985).
19. Minges, M. C., Cutback in Construction Funding Stirs Doubt on Water Quality Goals, *J. Water Pollut. Control Fed.* 45:199–201 (1973).
20. Piecuch, P. J., EPA Outlines PL92.500 Strategy, *J. Water Pollut. Control Fed.* 45:779 (1973).
21. Parkhurst, J. D., WPCF President's Message, *J. Water Pollut. Control Fed.* 46:2089–2092 (1974).
22. Sliter, J. T., Needed: $350 Billion — and a New Needs Survey, *J. Water Pollut. Control Fed.* 46:2254 (1974).
23. Landy, M. K., M. J. Roberts, and S. Thomas, *The Environmental Protection Agency: Asking the Wrong Questions,* New York: Oxford University Press (1990).
24. Marshall, E., Science Advisers Need Advice, *Science* 245(4913):20–22 (1989).
25. Jasanoff, S. S., Contested Boundaries in Policy-relevant Science, *Social Stud. Sci.* 17:195–230 (1987).
26. Yosie, T. F. and J. C. Kurtz, Peer-review Processes Used in Regulatory Decision Making, *Environ. Toxicol. Chem.* 6:491–493 (1987).
27. Reisa, J. J., Foreword, in *Methods for Ecological Toxicology: A Critical Review of Laboratory Multispecies Tests,* A. S. Hammons, J. M. Giddings, G. W. Suter, II, and L. W. Barnthouse, Eds., Ann Arbor, Michigan: Ann Arbor Science Publishers (1981), p. iii.
28. Duffus, J. H., *Environmental Toxicology,* London: Edward Arnold (1980).
29. van Genderen, H., Ecotoxicology: Fundamental and Policy Supporting, *Vakblad voor Biologen,* 65(13/14):2 (1985).
30. Moriarty, F., *Ecotoxicology: The Study of Pollutants in the Environment,* 2nd ed., London: Academic Press (1990).
31. Halffman, W., Mapping Environmental Sciences (1974–1989): Empirical Explorations and First Results, Unpublished paper, University of Amsterdam: Science Dynamics (1992).
32. Alkema, E., The Origins and Development of BION Work Communities in the Field of Ecology, Report, University of Amsterdam: Science Dynamics (1985).
33. Advisory Council for Science Policy [Raad van Advies voor het Wetenschapsbeleid (RAWB)], Advice on Ecotoxicology, Report, Amsterdam: BION (1987).
34. Letter on behalf of the BION Bureau by Tj. de Cock Buning to J. Dijkhof of NWO, April 15, 1988.
35. BION, Annual Report, Amsterdam (1990).
36. Truhaut, R., Ecotoxicology: Objectives, Principles and Perspectives, *Ecotoxicol. Environ. Saf.* 1:151–152 (1977).
37. Butler, G. C., Ed., *Principles of Ecotoxicology,* Chichester: John Wiley (1978).
38. Bourdeau, Ph., In Memoriam Jan G. P. M. Smeets, *Ecotoxicol. Environ. Saf.* 13:135–136 (1987).
39. Smeets, J., Ecotoxicological Aspects of the European Environmental Action Programs, *Ecotoxicol. Environ. Saf.* 2:43–150 (1978).
40. Smeets, J., New Challenges to Ecotoxicology, *Ecotoxicol. Environ. Saf.* 3:116–121 (1979).

41. Miller, D. R., General Considerations, in *Principles of Ecotoxicology*, G. C. Butler, Ed., Chichester: John Wiley (1978), pp. 3–17.
42. Moriarty, F., *Ecotoxicology: The Study of Pollutants in the Environment*, 2nd ed., London: Academic Press (1990).
43. Cairns, J., Jr., Will the Real Ecotoxicologist Please Stand Up?, *Environ. Toxicol. Chem.* 8:843–844 (1989).
44. Walton, B., Structure-activity Relationships in Environmental Toxicology and Chemistry, *Environ. Toxicol. Chem.* 7:403–404 (1988).
45. Desai, S. M., R. Govind, and H. H. Tabak, Development of Quantitative Structure-activity Relationships for Predicting Biodegradation Kinetics, *Environ. Toxicol. Chem.* 9:473–477 (1990).
46. Hermens, J., Effects of Complex Mixtures of Water Pollutants, *Vakblad voor Biologen,* 65(13/14):29–34 (1985).
47. Nirmalakhandan, N. and R. E. Speece, Structure-activity Relationships: Quantitative Techniques for Predicting the Behavior of Chemicals in the Ecosystem, *Environ. Sci. Technol.* 22(6):606–615 (1988).
48. Smies, M., F. Balk, J. Bok, and W. F. ten Berge, Environmental Toxicology in Industry, *Prof. J. Biol.* 56(13/14):87–90 (1985).
49. Latour, B., *Science in Action,* Milton Keynes: Open University Press (1987).
50. Cairns, J., Jr. and K. L. Dickson, A Simple Method for the Biological Assessment of the Effects of Waste Discharges on Aquatic Bottom-dwelling Organisms, *J. Water Pollut. Control Fed.* 43:755–772 (1971).
51. Taras, M., Standard Methods for the Examination of Water and Wastewater: Changes in the 13th Edition, *J. Water Pollut. Control Fed.* 43:150–152 (1971).
52. Anon., EPA Water Quality Criteria, *J. Water Pollut. Control Fed.* 44:712 (1972).
53. Resh, V. W. and J. D. Unzicker, Water Quality Monitoring and Aquatic Organisms: The Importance of Species Identification, *J. Water Pollut. Control Fed.* 47:9–19 (1975).
54. Anon., Research and the Quest for Clean Water, Report of the WPCF Research Committee, *J. Water Pollut. Control Fed.* 47:240–251 (1975).
55. Wall, T. W. and R. W. Hanmer, Biological Testing to Control Toxic Water Pollutants, *J. Water Pollut. Control Fed.* 59:7–12 (1987).
56. Cairns, J., Jr., Aquatic Field Testing: Experimental Mesocosm Field Techniques, *Reg. Toxicol. Pharmacol.* 8:226–238 (1988).
57. Cairns, J., Jr., Ed., *Aquatic Microbial Communities,* New York: Garland Publishing, Inc. (1977), p. 95.
58. Koeman, J. H., Ecotoxicological Evaluation: The Eco-side of the Problem, *Ecotoxicol. Environ. Saf.* 6:359 (1982).
59. Cairns, J., Jr., Biological Monitoring, Part IV: Future Needs, *Water Res.* 15:944 (1981).
60. Beyers, F. J., A Characteristic Diurnal Metabolic Pattern in Balanced Microcosms, *Publ. Inst. Mar. Sci., Univ. Tex.* 9:19–27 (1963).
61. Limoges, Ç., A. Cambrosio, D. Charron, and B. Longpré, From Ecotoxicology to Risk Assessment in Biotechnology: The Standardization of Microcosms, Paper for the Society of the History of Technology Conference in Uppsala, Sweden (1992).
62. Forbes, S. A., The Lake as a Microcosm, *Ill. Nat. Hist. Surv. Bull.* 15:537–550 (1987).

63. Odum, H. T. and C. M. Hoskins, Metabolism of a Laboratory-stream Micro-cosm, *Publ. Inst. Mar. Sci., Univ. Tex.* 4:115–133 (1957).

64. Debus, A. G., *Man and Nature in the Renaissance,* Cambridge Massachusetts: Cambridge University Press (1978).

65. Metcalff, R. L., G. K. Sangha, and I. P. Kapoor, A Model Ecosystem for the Evaluation of Pesticide Biodegradability and Ecological Magnification, *Environ. Sci. Technol.* 5(8):709–713 (1971).

66. Taub, F. B., A Continuous Gnotobiotic (Species-defined) Ecosystem, in *Aquatic Microbial Communities*, J. Cairns, Jr., Ed., New York: Garland Publishing (1977), pp. 105–138.

67. Kersting, K., Effects of Toxic Substances on Ecosystems, *Vakblad voor Biologen,* 65(13/14):118–119 (1985).

68. Cooke, D. G., Experimental Aquatic Laboratory Ecosystems and Communities, in *Aquatic Microbial Communities*, J. Cairns, Jr., Ed., New York: Garland Publishing (1977), pp. 59–103.

69. Giddings, J. M., Laboratory Tests for Chemical Effects on Aquatic Population Interactions and Ecosystem Properties, in *Methods for Ecological Toxicology: A Critical Review of Laboratory Multispecies Tests*, A. S. Hammons, J. M. Giddings, G. W. Suter, II, and L. W. Barnthouse, Eds., Ann Arbor, Michigan: Ann Arbor Science Publishers (1981), p. 59.

70. Cairns, J., Jr., *Multispecies Toxicity Testing,* New York: Pergamon Press (1985).

71. Cairns, J., Jr., *Community Toxicity Testing,* Philadelphia: American Society for Testing and Materials (1986).

72. Kimball, K. D. and S. A. Levin, Limitations of Laboratory Bioassays: The Need for Ecosystem-level Testing, *Bioscience* 35:167 (1985).

73. Odum, E. P., The Mesocosm, *Bioscience* 34(9):559 (1984).

74. Hammons, A. S., J. M. Giddings, G. W. Suter, II, and L. W. Barnthouse, Eds., *Methods for Ecological Toxicology: A Critical Review of Laboratory Multispecies Tests,* Ann Arbor, Michigan: Ann Arbor Science Publishers (1981).

75. Giesy, J. P., SETAC: Part of the Solution or Part of the Problem?, *Environ. Toxicol. Chem.* 9:1327–1330 (1990).

76. Perez, K. T. and G. E. Morrison, Environmental Assessment from Simple Test Systems and a Microcosm: Comparison of Monetary Costs, in *Multispecies Toxicity Testing*, J. Cairns, Jr., Ed., New York: Pergamon Press (1985), pp. 89–95.

77. Thomas, L. M., Strategies for Research in the U.S. Environmental Protection Agency, *Environ. Toxicol. Chem.* 8:273–275 (1989).

78. Jasanoff, S., Policy Shortfalls, *Science* 248:895–896 (1990).

79. Cortesi, R. S., EPA's Office of Exploratory Research, *Environ. Toxicol. Chem.* 7:91–93 (1988).

80. Cairns, J., Jr,. Putting the Eco in Ecotoxicology, *Reg. Toxicol. Pharmacol.* 8:226–238 (1988).

81. Fujimura, J., Constructing "Doable Problems" in Cancer Research: Articulating Alignment, *Social Stud. Sci.* 17:257–93 (1987).

CHAPTER 3

Endpoints of Interest at Different Levels of Biological Organization

Glenn W. Suter II

INTRODUCTION

DEFINITIONS

The term **endpoint** has two distinct uses in ecological assessment: measurement endpoint and assessment endpoint.[1,2] Assessment endpoints are explicit definitions of the environmental values that are to be protected by an assessment. Depending on the specificity of the assessment, they may be very broadly and qualitatively defined (e.g., no loss of species) or very specifically and quantitatively defined (no more than one episode per 5 years in which ten or more fish are killed by an effluent release). Measurement endpoints are the output of a toxicity test, series of field measurements, or other attempts to measure the responses of a biological system to a toxicant or other agent. Examples include 96-h LC50s, 7-d NOECs, mean Secchi depth, and catch per unit effort. Measurement endpoints for toxicity tests are referred to as **test endpoints**. Because the topic of this volume is toxicity testing, the remainder of the discussion will be limited to test endpoints and their relation to assessment endpoints.

The point of distinguishing test endpoints from assessment endpoints is to highlight the following three facts. First, although test endpoints are often used in assessments as *de facto* assessment endpoints, the two types of endpoints seldom correspond. Assessment endpoints nearly always refer to effects on higher levels of ecological organization, larger spatial scales, and longer time periods than even the largest mesocosm test, and nearly always include more or different species, life-stages, and ecosystem types than toxicity tests. Therefore, some extrapolation from test endpoint to assessment endpoint is needed. For example, if we wish to prevent loss of 5% or more of the species in a fish community due to continuous release of a chemical, and our only test endpoint for that chemical is an LC50 for juvenile fathead minnows, we must extrapolate to the full set of species, life stages, exposure patterns, and relevant responses.

Second, test endpoints are not natural features of toxicity tests, and standard test endpoints are not necessarily the best estimators of assessment endpoints. Toxicity tests are complex biological and chemical events from which a virtually infinite number of observations could be taken and from which numerous observations usually are taken, but all observations are boiled down to a standard test endpoint. For example, the periodic counts of dead fish conducted during a 96-h lethality test could be used to calculate the conventional test endpoint (the 96-h LC50), a median time to death (LT50), an LC10, a concentration-response function for lethality, a concentration-duration function for 50% lethality, or a concentration-duration-response function for lethality. In addition, numerous nonlethal responses could be observed and used to determine additional endpoints from the same test such as time to first appearance of distress symptoms.

Third, automatic adoption of a standard test endpoint implies adoption of a standard test design, which will not necessarily allow the most useful observations to be made. If the decision is how best to release a particular volume of wastewater, the testing effort might best be directed to testing the effects of varying the release rate (i.e, comparison of high concentration-short duration exposures to lower concentrations-longer duration exposures) rather than varying only concentration as in a 96-h LC50. Unless constrained by regulations to generate a standard test endpoint, one should decide which test design and endpoint or set of test endpoints could best be used to evaluate the assessment endpoint.

CHOOSING TEST ENDPOINTS

In 1977, Sprague and Fogels suggested that more attention should be paid to why toxicity tests are being performed.[3] This sage advice has been largely ignored as ecological toxicologists argue about appropriate species, life stages, durations, levels of organization, etc. Tests and test endpoints should be chosen to serve environmental management goals and purposes. There are two fundamental purposes to which test endpoints may be applied.

The first purpose is prediction. Most tests are conducted for the purpose of supporting a prediction concerning the effects of releasing the tested chemical. In the simplest case, this purpose is satisfied by a screening assessment that classifies the contaminant into one of two or more categories. For example, at a waste site, the ecological risk assessment often must begin by screening a hundred or more chemicals into those that are worthy of further analysis, testing, and modeling (i.e., contaminants of concern), and those that can be ignored. Regulatory criteria and standards are also developed for screening. For example, the national ambient water quality criteria screen chemical concentrations in ambient waters into acceptable or unacceptable categories. In screening assessments, the prediction is that chemical exposures below the screening benchmark values will not cause unacceptable effects (however defined). At the other extreme, some predictive assessments must actually estimate the nature and level of effects. For example, in the case of ecological assessments for waste sites where damage to a receiving stream is documented, it is necessary to predict the benefits to the stream ecosystem of alternative remedial actions so that they can be compared to each other, to the costs of the alternative remedial actions, and to the amount of ecological damage done by the remedial actions due to capping, dredging, filling, etc. In general, predictions of the nature and magnitude of effects are needed when alternative actions must be compared, when costs and benefits must be balanced, and when monetary damages must be assessed for ecological injuries.

In practice, there is a continuum of predictive assessments from crude screens to quantitative estimates. When the costs of conservatism are relatively low, the need for predictive power is also relatively low. That is, one can use a few cheaply obtained test endpoints that are not good predictors of the nature or magnitude of effects if, with the right safety factors, they can give a screening benchmark that is reliably protective. Quantitative structure activity relationships (QSARs) might even be used in place of testing. In the case of screening potential contaminants of concern, the cost of a false positive (i.e., including chemicals as contaminants of concern when they are, in fact, innocuous at the waste site) in terms of additional chemical analyses, etc. may be small relative to the cost of doing enough testing to reduce significantly the number of false positives. On the other hand, both the economic costs of overly conservative national water quality criteria and the ecological impacts of underconservative national criteria could be quite large. Therefore, it could be worthwhile to spend enough on testing to allow good predictions to be made of the nature and magnitude of effects associated with different criteria levels for common criterion chemicals.

When choosing test designs and endpoints for predictive assessments, it is important to remember that the test endpoint is only one input to the assessment. The other major input is the model used to extrapolate from the test endpoint to the assessment endpoint. The simplest model is that the test and assessment endpoints are identical. For example, the assessment endpoint may be prevention of large fish kills, and the 96-h LC50 for fathead minnows may be assumed to correspond to a threshold for occurrence of such kills. Other types of extrapolation models include safety factors, statistical models, and mechanistic mathematical models.[4–6] Hence, the best test endpoint is not necessarily the one that most closely corresponds to the assessment endpoint, but rather, it is the one that, in combination with an extrapolation model, provides the best estimate of the assessment endpoint.

The second purpose of toxicity testing is to support ecological epidemiology by providing the evidence necessary to interpret field studies and prove causation. Field studies can establish the occurrence and magnitude of effects but cannot prove causation because of the potential for confounding variables in uncontrolled studies to create false associations or obscure true associations of causes and effects. Causation can be established by demonstrating the consistency of the field results with those of controlled studies (i.e., toxicity tests).[7] Tests designed for predictive assessments may serve this purpose. For example, if there are no fish in a stream that constitutes suitable habitat and the concentration of a chemical in a stream greatly exceeds the LC50 for that chemical, then one can be reasonably confident that the chemical is a cause. However, in many cases, the predictions of effects derived from the standard test endpoints are uncertain, and the effects observed in the field could be caused by any of a number of factors. In such cases, diagnostic test endpoints are needed. These would be test endpoints that are indicative of a particular contaminant or class of contaminants. These might include body burdens of the contaminants, frequencies of histopathologies, and levels of biochemical biomarkers. This purpose of toxicity testing has been relatively neglected.

LEVELS OF ORGANIZATION

A few general points need to be made about hierarchical systems before discussing endpoints for the different levels of organization.[8] First, the dynamics of entities at one level of organization constitute the mechanisms that determine the dynamics at the next higher level of organization. For example, the responses of individual organisms are the mechanisms by which the response of a population is generated. Death or infertility of organisms generate population declines. Second, the dynamics of one level of organization are constrained by the next higher level of organization. For example, the size of a population is constrained by the productivity of the ecosystem that it inhabits.

Mechanistic science investigates lower-level phenomena to explain or predict phenomena at the level of interest. This tendency is reflected in the recent popularity of toxicodynamic models to explain the effects of chemicals on organisms and populations, and of individual-based models to explain the effects of various disturbances on populations and communities. However, the constraints imposed by higher levels may render mechanistic models moot. For example, if algal production is limited by nutrients, it would be futile to predict toxic effects on the primary productivity of a lake on the basis of the reduction in photosynthetic capacity of individual algal populations, without knowledge of those constraints.

The implication of these characteristics of hierarchies is that selection of test endpoints depends on the nature of the assessment problem and on available knowledge of the systems being assessed. If one chooses test endpoints at the same level of organization as the assessment endpoint, the empirical basis for the assessment is strong, but there is little basis for extrapolation. For example, a fathead minnow acute lethality test can be assumed to represent the mortality of all individual fish in the field with some error due to variance in species sensitivities, etc. However, neither the mechanisms by which the toxic effects are imposed nor the constraints imposed by the larger system are known. If test endpoints address lower levels of organization, then mechanisms of action can be identified and used to simulate the higher-level effects. This approach potentially provides great predictive power. For example, if one can assume that toxic effects of narcotic chemicals on fish are strictly a function of passive uptake and partitioning, then the effects of a new chemical on various fish species can be accurately predicted from ventilation and the relative volume of major internal compartments.[9] However, if other mechanisms are, in fact, involved in the imposition of effects such as specific modes of toxic action, active metabolism, or avoidance, then the mechanistic model will not predict effects on even one species. If tests are performed at higher levels of organization than the assessment endpoints, then constraints may be identified. For example, the resistance of a fish population to a chemical's toxic effects may be constrained by a lack of food if zooplankters are sensitive to the chemical. However, the constraints operating in a small model ecosystem may be very different from those in a real-world ecosystem. In a real lake, resistant zooplankters may replace sensitive ones, or recolonization and recovery of zooplankter populations may be rapid. Constraints also vary among real world systems. Hence, knowledge of constraints, like knowledge of mechanisms, may be either revealing or misleading.

SUBORGANISMAL ENDPOINTS

Suborganismal endpoints include biochemical, physiological, and histological characteristics. Measurements of suborganismal characteristics for the purpose of identifying exposure to, or effects of, chemicals or other hazardous agents are commonly referred to as biomarkers. Assessment endpoints are not defined in terms of suborganismal states because those states are not of interest to environmental managers. That is not to say that they are not important. Rather, they are important in so far as they serve to predict future states of higher level systems (organisms, populations, and ecosystems) or serve to indicate the current state of such systems.

Suborganismal test endpoints are currently little used in predictive assessments because we do not have models to extrapolate from suborganismal responses in tests to higher-level responses in the field. This situation is not due to a lack of attention to development of assessment models but rather is primarily due to a lack of knowledge of what particular biochemical or histological changes imply for the fate of organisms.

Suborganismal test endpoints are much more useful in ecological epidemiology. The most conspicuous example is the cholinesterase inhibition. Toxicity tests have established the range of brain and blood cholinesterase levels at which various vertebrate taxa die, and these can be used to determine whether dead or moribund organisms discovered in the field are victims of organophosphate cholinesterase-inhibiting pesticides.[10] Research is needed to develop other diagnostic biomarkers or sets of biomarkers.

ORGANISMAL ENDPOINTS

Organismal endpoints include death, life span, growth rate, fecundity, deformity, and various behaviors. Organismal assessment endpoints are controversial. In many cases, pollutants can be shown to kill or otherwise injure individual organisms without any apparent consequences at the population or ecosystem level. Mortality rates in nonhuman populations are generally quite high and compensatory mechanisms at the population and ecosystem levels can readily absorb a few more mortalities. It is the abundance or productivity of the populations that determine their utilitarian and nonutilitarian value, not the fate of the component individuals. Hence, from an ecological perspective, organismal assessment endpoints are of little interest. In addition, the harvesting of individual plants and animals for food, materials, and recreation suggests that the continued existence of individual nonhuman organisms is not, in general, a societal concern. However, it is often easier to demonstrate an organismal effect than a higher-level effect, and agencies have taken regulatory actions on the basis of evidence of organismal effects. The goal of the U.S. National Ambient Water Quality Criteria has been described as "protect 99% of individuals of 95% of species".[11]

Most standard toxicity tests are conducted at the organism level. The most common standard endpoints are the median lethal dose or concentration (LD50 or LC50) and the thresholds for statistically significant changes in survival, growth, and fecundity (NOELs and LOELs). Alternative endpoints would be more useful for assessment. For example, chronic toxicity to early life stages of fish might be better represented by a percent reduction in the weight of juveniles per initial egg than by NOELs or LOELs, which correspond to no particular level or type of effect. Factors, statistical models, and mathematical models have been developed to extrapolate organism-level test endpoints to higher levels of organization and between different species and age classes of organisms.[4-6]

Organism-level test endpoints are commonly used in ecological epidemiology as well as predictive assessments. Because organism-level test endpoints are by far the most common toxicity data available, the cause of effects in the field are commonly evaluated by comparing concentrations in water or other ambient media to LC50s, LOECs, etc. However, organism-level responses are generally less chemical specific and therefore less diagnostic than suborganismal

Table 1 Generic Population Endpoints Recommended by a SECOTOX Workshop[12]

Assessment Endpoints

1. No kills: Incidents in which large numbers of organisms are killed are generally recognized to be undesirable
2. No significant reductions in productivity: This endpoint is most appropriate to populations of resource species that are harvested
3. No significant reductions in abundance: This endpoint is more appropriate to nonresource species; it is easier to measure than productivity and is more apparent to nonconsumptive users such as bird watchers and hikers
4. No significant reduction in population range: This endpoint is useful for regional scale assessments in which range reductions are readily related to the geographic distribution of habitats, disturbances, and contaminant sources
5. No extinctions: This is particularly relevant to assessment of rare or endangered species
6. No significant loss of genetic diversity: This endpoint is associated with the ability of a population to resist and recover from perturbations
7. No significant loss of population quality: Quality is a rather broad term that includes contamination of a population of human food species and high frequencies of tumors, lesions, and deformities that make organisms repellant

Measurement Endpoints

1. Mortality, age specific
2. Fecundity
3. Growth

responses. Organism-level endpoints are particularly useful when lethality is involved because the number of agents that would kill a large number of organisms at a particular place and time is likely to be small. In the more common case where a population is found to be less abundant or productive than expected, many causes may be responsible and organism-level test endpoints are useful but less conclusive evidence of causation than in the case of kills.

POPULATION ENDPOINTS

A recent SECOTOX workshop discussed the subject of population endpoints (Table 1).[12] Consensus was readily reached on population-level assessment endpoints because the commercial, recreational, and aesthetic values of populations are apparent. The test endpoints on which the participants could agree are the traditional demographic parameters: life-stage-specific mortality, growth, and fecundity rates. As noted above, these parameters are measured in organism-level tests. Population models developed by resource managers such as Leslie matrices can readily convert changes in these parameters into predictions of production or abundance for predictive assessments.[5]

Toxicity tests are seldom designed to generate population-level endpoints. The major exception is algal toxicity tests for which the test endpoints are growth rate, biomass, or abundance of an algal population in a flask. Multigenerational tests of zooplankters can also generate population-level endpoints such as the intrinsic rate of increase (r).[5] However, they more commonly use the statistical significance of the difference of individual parameters from controls as test endpoints. Finally, many of the test endpoints of ecosystem-level tests are population-level parameters. In particular, most microcosm and mesocosm tests report abundances of component populations.

Population-level test endpoints are even less diagnostic of particular chemicals than organism-level endpoints. However, it is often easier to relate population-level test endpoints to field observations than organism test endpoints. For example, if crustacean zooplankter species are rare or missing from a lake, the fact that the average exposure concentration of a chemical measured in the lake causes a negative population growth rate in laboratory daphnid tests obviously supports the inference that the chemical is the cause. A traditional organism-level test endpoint (e.g., the number of young per brood is statistically significantly less than controls) is not so readily interpreted relative to the field observations because it does not include other components of the population response (e.g., longevity) and does not imply any particular level of effect on the response that is included.[4,5]

ECOSYSTEM ENDPOINTS

Potential ecosystem-level assessment endpoints include structural properties such as species number, biomass, and trophic structure, and functional properties such as production and respiration. Each of these can be measured in an entire system such as a lake, in a structural component such as the pelagic and benthic communities, and in taxonomic or functional groups such as fish and phytoplankton. Ecosystem assessment endpoints are more problematical than those at other levels. The number of ecosystem attributes that might constitute assessment endpoints is large, and none of them represent compelling goals of ecosystem management in the same way that production, abundance, and range represent the goals of population management. Although nearly everyone agrees that ecosystem protection is a desirable goal, divergence occurs when operational definitions must be formulated. Preservation of a natural state might seem to be the goal around which ecosystem assessment endpoints could be formulated, but few ecosystems are not managed in some way, and even systems that are undisturbed do not assume an identifiable natural state.[13] Even national parks spray insect pests, clear balds, control fires, and otherwise interfere in natural processes to maintain ecosystems in an acceptable state. As a result, the goals of ecosystem management are ad hoc and are often vaguely defined.

Ecosystem test endpoints are equally problematical. Most of the endpoints of microcosm and mesocosm tests are attributes of populations (e.g., abundance) and organisms (e.g., mortality). U.S. Environmental Protection Agency protocols for the six terrestrial and aquatic microcosm toxicity tests include the following classes of ecosystem parameters: net production, net respiration, chlorophyll a content, glucose decomposition rate, nutrient content, nutrient loss rate, dissolved oxygen, alkalinity, conductivity, and abundance or diversity of various groups of species (e.g., algal abundance).[14] The test endpoint in each case is a statistically significant deviation from the controls in any of the large number of measured ecosystem, population, and organism-level parameters. This suggests that a deviation of any type or magnitude in an ecosystem is unacceptable. A more useful alternative would be to calculate exposure-response curves and percent reductions in specific ecosystem functional and structural properties.[15]

Recently, there seem to be signs of movement toward use of species richness as a standard ecosystem test and assessment endpoint. Pratt found that it is most often the most sensitive response measured in ecosystem-level tests.[16] Dickson et al. found that, in several studies of streams and rivers subject to toxic effluents, species richness was more consistently affected by toxicity than other measured parameters.[17] The use of species richness as an assessment endpoint is consistent with the recent increase in concern with maintenance of biodiversity.[18]

A general strategy for selecting ecosystem assessment endpoints is to focus on management goals for particular ecosystem types that are expressed as ecosystem properties. There are two obvious examples from the management of lakes: (1) avoid accelerated eutrophication; (2) maximize the proportion of primary production that is transfered to pelagic predatory fish (i.e., game fish).

The use of ecosystem test endpoints in predictive assessments has been simplistic. It has been assumed that any ecosystem test endpoint corresponds to all ecosystem assessment endpoints. Stated in that way, the assumption seems naive. It is unlikely that all ecosystems and all ecosystem responses would be equally sensitive, particularly when some of them are in jars in laboratories. However, in the author's experience, regulatory agencies have not applied safety factors or other extrapolation models to ecosystem test endpoints when performing assessments of chemicals.

Ecosystem test endpoints have seldom, if ever, been used to interpret causation in ecological epidemiology. The responses of ecosystem functional parameters such as production and respiration are unlikely to be diagnostic of particular chemicals or chemical classes. However, changes in the relative abundances of species or other taxa observed in an ecosystem-level test may be diagnostic due to consistent differences in the relative sensitivities of species to particular chemicals. This use of ecosystem tests would be an extension of the indicator species concept. For example, increases in rotifer abundance relative to crustacean zooplankton may be indicative of a number of toxic chemicals.

GENERALIZATIONS

In general, entities at higher levels of organization are more important. We do not care about loss of an enzyme molecule, care little about loss of an organism, care considerably more about loss of a population, and care most of all about loss of an entire ecosystem. However, that does not mean that all ecosystem properties are more important than all properties of lower level entities. For example, we are likely to care a great deal more about a decline in productivity of a game fish population than about a change in phytoplankton community evenness.

In general, higher level entities are more diverse in their responses than lower level entities. One acetylcholinesterase molecule responds like another to an organophosphate pesticide. Although individuals of different fish species are much more different than their constituent biochemicals, it is a truism of ecological toxicology that a fish is a fish.[3] At the other extreme, ecosystems incorporate all of the variation in sensitivity among organisms and populations plus variation in ecosystem-level structural and functional properties. Hence, it is doubtful that responses of even a relatively complex ecosystem test like a pond mesocosm can, in fact, predict the responses of lakes, rivers, and streams in the sense that fathead minnow responses predict the responses of bluegill, rainbow trout, and Tennessee dace. It seems even less likely that the responses of lakes, rivers, and streams will be predicted by responses of a simple ecosystem test system such as the protozoan community on sponge blocks.[19]

Properties of ecosystems and other higher level entities tend to be more predictable in the sense that they are more resistant to change. Because of functional redundancy, ecosystems tend to maintain functions such as production and respiration and structural properties such as trophic structure because declining or disappearing sensitive species are replaced by increasing or colonizing resistant species. To a lesser extent, populations are buffered from effects on individuals by density-dependent compensatory mechanisms, which also contribute to the stability of ecosystems. To an even lesser extent, organisms are protected by homeostatic mechanisms, which also contribute to the stability of populations and ecosystems. In other words, if one visited a river reach over a period of several years that was subject to a variety of perturbations, it is likely that the original basic structure and function of the ecosystem would still be present, it is likely that many but not all original populations would be present, but it is unlikely that any of the original organisms would be present. However, if the basic structure and function of the system is changed (e.g., the river is dammed) that change would imply significant changes in the constituent populations and individuals.

In general, testing becomes more difficult and modeling becomes more necessary at higher levels of organization. In large part this is because the spatial scale and time to response become larger at larger scales. It has been suggested that this problem can be avoided by testing microbial ecosystems, but there is no reason to expect that their responses would be predictive of responses of macroecosystems.[19] It is not even clear that simple statistical extrapolation models like those used to extrapolate organism-level test endpoints between species can be devised to extrapolate from practical ecosystem test endpoints to real ecosystems. At regional and global levels, toxicity testing is unacceptable and practically impossible. When mechanistic models are used, the organizational level of the mechanisms must coincide with the level of the test endpoints used to paramaterize the model. Currently the most practical approach is to use the demographic mechanisms (fecundity, death, and growth) that are generated by the current standard toxicity tests. However, at population and higher levels of ecological organization, "individual-based" models are increasingly popular.[20] Use of such models for ecotoxicological risk assessments requires either that test endpoints include the behavioral and physiological parameters that drive such models or that the assessor make assumptions about the mode of action of the chemical with respect to those parameters.

CONCLUSIONS

Ecotoxicologists have mistakenly assumed that test designs and endpoints can be selected strictly on the basis of their inherent properties such as cost, time required, realism, variance among replicates, and level of organization. These considerations are important, but they should be secondary to the question, "What test endpoint or set of test endpoints will provide the best basis for an assessment that will lead to an informed decision?" Hence, the choice should not be made by the toxicologist alone, but rather the toxicologists must collaborate with assessors and both must be guided by the information needs of the environmental risk manager and his or her understanding of the public values involved in the decision. The risk manager can be educated by the toxicologist and assessor as to the ecological importance of alternative assessment endpoints. However, they cannot substitute their judgment for his or hers, because he or she is the designated representative of the public, which will pay for the consequences of regulation or suffer the loss of resources. This approach makes test endpoint selection a component of a process of environmental decisionmaking to which toxicologists must contribute rather than a component of the science of toxicology.

Selection of tests for decisionmaking requires answers to the following questions.

1. What is the nature of the decision?
2. What societal values are involved in the hazard to be assessed?
3. How can those values be operationally defined as assessment endpoints?
4. What combination of models, test endpoints, and other data will most efficiently provide an assessment of the assessment endpoints in a form suitable for the decision?

The answers to these questions will differ for predictive vs. retrospective assessments, for screening vs. estimation, for site-specific vs. national assessments, and among nations and generations.

This approach is reasonable when the time and resources are available to perform toxicity testing ad hoc. However, most assessments must rely on existing toxicity data, most of which are in the form of a few standard test endpoints. Given that tests are being conducted with new chemicals and materials or new species or test systems, and those results expressed as test endpoints will be used in diverse assessments, what test endpoints are best for general use? When assessments are performed using existing toxicity data, the toxicologist is not directly involved, so the tailoring of the assessment to the decision must be performed primarily by the assessor. The toxicologist then should provide the assessor with greater flexibility, which means providing more information about the outcome of toxicity tests.

This flexibility might best be attained by reporting a series of test endpoints for each test. Consider again the example of standard acute fish lethality tests, which conventionally generate as an endpoint the 96-h LC50. That standard endpoint should always be reported because many existing assessment models employ LC50s and because it provides a basis of comparison to previously tested chemicals. However, many assessors would like to estimate concentrations causing proportional mortalities other than 0.5 so the probit function or other model used to generate the LC50 should be presented as a second test endpoint. In addition, most protocols for acute lethality tests require that mortalities be reported at intermediate intervals (i.e., 24, 48, 72, and 96 h), so a probit plane or other concentration-time-mortality model might be presented as a third endpoint. This would allow assessors to address effects of exposure durations other than 96-h. If a 3-D model could not be fit to the data, the same information could be presented as a concentration by duration matrix in which the cells would contain the observed proportional mortality. Finally, the original data should be reported so that the assessor could calculate alternative benchmarks. The test results could also be made more useful for ecological epidemiology if behaviors, pathologies, and biomarkers associated with lethality were reported along with the endpoints intended for predictive assessments.

In sum, the problem of what level of organization to use in toxicity testing has a simple answer. Use the tests that generate the test endpoints that can best support the goals of the assessment. In general, this will be the test endpoint that, in combination with an extrapolation model, provides the best estimate of the assessment endpoint. However, other goals such as diagnosing the cause of ecological injuries require other test endpoints. You cannot choose the best means of travel until you know where you want to go.

The difficulty in this apparently simple approach is obtaining a clear statement of environmental goals that can be operationally defined. Often, regulatory agencies are reluctant to clarify the basis for their decisions. However, the use of fifth percentiles of species sensitivity distributions as regulatory assessment endpoints in the United States and Europe constitutes a generic environmental goal.[21, 22] That is, at least 95% of species in an ecosystem exposed to a hazardous agent should escape injury. If no such goal is available or forthcoming, ecological toxicologists and risk assessors must define their assessment endpoints as clearly as possible and attempt to obtain the concurrence of the risk manager. From the precedents provided by such actions, additional operationally defined environmental goals will emerge.

ACKNOWLEDGMENTS

This manuscript benefited from the comments of Ruth Hull, Bob O'Neill, and two anonymous reviewers. It is publication No. 4058, of the Environmental Sciences Division, ORNL, which is managed by Martin Marietta Energy Systems, Inc., under contract DE-AC05–84OR21400 with the U.S. Department of Energy.

REFERENCES

1. Suter, G. W., II, Ecological End Points, in *Ecological Assessment of Hazardous Waste Sites: A Field and Laboratory Reference Document*, W. Warren-Hicks, B. R. Parkhurst, and S. S. Baker, Jr., Eds., EPA 600/3–89/013, Corvallis Environmental Research Laboratory, Oregon (1989), pp. 2–1-2–28.
2. Risk Assessment Forum, *Framework for Ecological Risk Assessment*, EPA/630/R-92/001, U.S. Environmental Protection Agency, Washington, D.C. (1992).
3. Sprague, J. B. and A. Fogels, Watch the Y in Bioassay, in *Proceedings, 3rd Aquatic Toxicity Workshop*, Environmental Protection Service Technical Report No. EPS-5-AR-77–1, Halifax, Canada (1977), pp. 107–115.
4. Suter, G. W., II, Organism Level Effects, in *Ecological Risk Assessment*, G. W. Suter, II, Ed., Boca Raton, FL: Lewis Publishers (1992), pp. 175–246.
5. Barnthouse, L. W., Population Level Effects, in *Ecological Risk Assessment*, G. W. Suter, II, Ed., Boca Raton, FL: Lewis Publishers (1992), pp. 247–274.
6. Suter, G. W., II and S. G. Bartell, Ecosystem Level Effects, in *Ecological Risk Assessment*, G. W. Suter, II, Ed., Boca Raton, FL: Lewis Publishers (1992), pp. 275–308.
7. Suter, G. W., II., Retrospective Ecological Risk Assessment, in *Ecological Risk Assessment*, G. W. Suter, II, Ed., Boca Raton, FL: Lewis Publishers 1992), pp. 311–364.
8. O'Neill, R. V., D. L. DeAngelis, J. B. Waide, and T. F. H. Allen, *A Hierarchical Concept of Ecosystems*. Monograph in Population Biology 23. Princeton University Press, Princeton, NJ (1986).
9. Lassiter, R. R. and T. G. Hallam, Survival of the Fattest: Implications for Acute Effects of Lipophilic Chemicals in Aquatic Populations, *Environ. Toxicol. Chem.* 9:585–596 (1990).
10. Fite, E. C., L. W. Turner, N. J. Cook, and C. Stunkard, Guidance Document for Conducting Terrestrial Field Studies, U. S. Environmental Protection Agency, Washington, D.C. (1988).
11. April, R. W., U.S. EPA Approaches to Ecological Risk Assessment, paper presented at the Chemical Manufacturers Association's Issues in Ecological Risk Assessment Workshop, Washington, D.C., April 7, 1992.

12. Suter, G. W., II and M. H. Donker, Workshop 10, Parameters for Population Effects of Chemicals, *Sci. Total Environ.,* Suppl., 1793–1797, 1993.
13. Botkin, D. B., *Discordant Harmonies,* Oxford University Press, New York (1990).
14. U.S. Environmental Protection Agency, Toxic Substances Control Act Test Guidelines: Proposed Rules. Fed. Reg. 52:36334–36371 (1987).
15. Babich, H., D. L. Davis, and J. Trauberman, Environmental Quality Criteria: Some Considerations, *Environ. Manage.* 5(3):191–205, 1981.
16. Pratt, J. R., Aquatic Community Responses to Stress: Prediction and Detection of Adverse Effects, in *Aquatic Toxicology and Risk Assessment, Thirteenth Volume,* W. G. Landis and W. H. van der Schalie, Eds., Philadelphia, PA: American Society for Testing and Materials (1990), pp. 16–26.
17. Dickson, K. L., W. T. Waller, J. H. Kennedy, and L. P. Ammann, Assessing the Relationship Between Ambient Toxicity and Instream Biological Response, *Environ. Toxicol. Chem.* 11:1307–1322, 1992.
18. Wilson, E. O., *Biodiversity,* Washington, D.C.: National Academy Press (1988).
19. McCormick, P. V., J. R. Pratt, and J. Cairns, Jr., Effect of 3-Trifluoromethyl-4-Nitrophenol on the Structure and Function of Protozoan Communities Established on Artificial Substrates, in *Community Toxicity Testing,* J. Cairns, Jr., Ed., Philadelphia, PA: American Society for Testing and Materials (1986), pp. 224–240.
20. Houston, M., D. DeAngelis, and W. Post, New Computer Models Unify Ecological Theory, *BioScience* 38:682–691, 1988.
21. Stephan, C. E., D. I. Mount, D. J. Hanson, J. H. Gentile, G. A. Chapman, and W. A. Brungs, Guidelines for Deriving Numeric National Water Quality Criteria for the Protection of Aquatic Organisms and Their Uses, PB85–227049, U.S. Environmental Protection Agency, Duluth, MN (1985).
22. Van Leeuwen, K., Ecotoxicological Effects Assessment in The Netherlands, *Environ. Manage.* 14:779–792 (1990).

CHAPTER 4

Role and Significance of Scale to Ecotoxicology

Kenneth T. Perez

INTRODUCTION

Physical models of systems have been used since the beginning of science. They are applied when a basic mechanistic understanding of the desired system is either unknown or restricted to some previously tested conditions, and the dynamic behavior under a new set of conditions is unpredictable. Physical models are used when the best mathematical constructs of the system under study are not able to provide sufficient predictive or descriptive power.

In the nonbiological disciplines, physical models or small-size replicas of a larger system predict dynamic behavior under various perturbation conditions. Disciplines such as aeronautics,[1] naval architecture,[2] and hydrodynamics[3] have used small-scale models. Examples include model aircraft as small as 1:2000 of actual scale, which have been used in wind tunnel simulations. Model ship hulls, 1:6 to 1:3 actual size, have been tested in tow tanks to determine designs that minimize frictional drag. Three-dimensional physical models of coastal environments have been used to predict localized flood conditions. In these cases, the model systems and the actual system are sufficiently similar in structure to make realistic extrapolations.

In the biological discipline of experimental medicine, the "model" for the human system is an animal system. Some animals are more appropriate models for given human diseases than others because of differences in size and age, as well as differences in the type (i.e., species, sex within species, etc.) of test animal chosen. For example, pigs are appropriate human analogues for the study of atherosclerosis compared to other animal models.[4] Some models give misleading results. Swenberg[5] has shown that male rats, because of their unique ability to induce alpha$_{2u}$-globulin when exposed to specific chemicals, inaccurately predict renal cancers in humans. Additionally, therapies that were effective for arterial injury restenosis following balloon angioplasty in rats, pigs, and rabbits failed in human studies.[6] These experiences demonstrate the need for caution in choice of model systems as well as interpretation of associated data.

Ecotoxicology is a discipline that "... predicts the impacts of chemicals upon ecosystems."[7] One of the methods used in ecotoxicology is development of a physical model of a specific environment. Predictions concerning the fate and effects of a test chemical are based upon results obtained from the model. These models have an assortment of names: microcosms, mesocosms, macrocosms, environmental simulators. The common feature of these models is that they usually contain more than one species of multicellular organism. The justification for using these models is similar to that in other disciplines. Specifically, either the costs for direct experimentation on a natural ecosystem are high, or the natural ecosystem is unavailable[8] because of the perceived value of the resource. There are some major difficulties in the use of physical models for ecotoxicological studies. First, definition of the physical limits of an ecosystem or even a habitat within an ecosystem can be difficult. In aeronautics, for example, it is easy to delineate the boundaries of an aircraft. Boundaries or limits of an ecosystem (and habitat) are not as easily identified. Second, isolation of an environment can result in artifacts caused by the physical containment devices. Third, variations in spatial and temporal scales of natural systems result in variable species composition and associated interdependencies that do not exist for objects such as planes and ships.

Do these difficulties preclude the successful application of physical models to ecotoxicology? Is it possible to apply physical environmental models in a way that their use can be as successful as in other disciplines? To address these questions, I shall outline the major issues of scale for physical models, when and how they are used to predict the fate and ecological effect of anthropogenic materials on natural ecosystems. The major scalar issues are (1) spatial scale, (2) temporal scale, and (3) simulation conditions.

This chapter will define the degree to which these issues have been addressed by previous studies and make some associated recommendations for future research.

CHARACTERIZATION OF NATURAL SYSTEMS

As a first principle, the development of any model implicitly assumes a sufficient knowledge of some true system being simulated. For environmental models, a quantitative definition in both time and space of the particular part, segment, or whole ecosystem is, therefore, necessary. Such information is needed for the following:

1. Identification of relevant variables of system state
2. Incorporation and reproduction of significant physical conditions or forcing functions (e.g., water exchange rates in aquatic systems, precipitation rates in terrestrial systems)
3. Equating physical model results to the natural system being simulated

Given this "principle", it is surprising how many studies using environmental models have neither measured nor defined the natural system they are attempting to simulate. Gearing[9] reviewed the literature on the use of microcosms for ecotoxicological studies of marine and freshwater systems. Of the 339 papers cited as examples of environmental types of model systems (marine, lacustrine, and streams), only 41, by Gearing's count (~12%) actually made field measurements. Examination of these articles revealed that simultaneous measurements in the physical model and natural system were made in only 29 of the 41 papers. The point is that fewer than 9% of these 339 papers considered and collected appropriate field data to design, develop, and verify the efficacy of their models.

It is clear that the utility of physical environmental models rests with the ability to define the degree to which these systems can dynamically mimic some behavior or property of the natural system from which they were derived. However, Menzel[10] argues that it is more important that enclosures closely duplicate themselves rather than the outside world. The value of a model, however, resides in its ability to mimic some real system, not itself. Accuracy or fidelity with the field is the first priority; precision, while important, is a secondary consideration. Characterization of the spatial and temporal behavior of the field system is essential for comparisons with physical model systems. Researchers in Sweden have fully embraced this requisite concept for their ecotoxicological studies[11] and have made statistical comparisons between their physical models and the field. In North American studies, Morton and colleagues[12] tested for statistical differences in structural and functional properties of their intact seagrass models with the natural seagrass system. Other investigators do not always perform rigorous statistical comparisons between laboratory and field systems. However, this omission did not prevent these investigators from making quantitative statements about observed numerical differences in the two systems. Phrases such as "surprisingly similar"[13] and "... behaved in a manner very similar to the waters outside the containers"[14] and "similar patterns"[10] and "qualitatively similar"[15] and "is significantly different from"[16] and "observations outside the enclosures…showed that extreme conditions did not occur in the natural environment"[17] were made without performing a statistical test to establish whether these stated differences, in fact, existed. Standard analysis of variance procedures, such as repeated measures analysis,[18] are essential for establishing differences between laboratory and field systems for the unperturbed or control treatment condition of any ecotoxicological study.

SPATIAL SCALE OF PHYSICAL MODELS

The dependence of species richness and abundances on spatial size is one of the major elements of scale that has influenced the way in which ecotoxicologists have designed and interpreted their physical models. The two major effects of spatial scale on physical models are (1) species inclusion or exclusion and (2) potential artifacts associated with containerization.

Seventy years ago, Gleason[19] showed that the number of species varied directly as a function of sample size. This simple relationship apparently has had an effect on the fledgling science of ecotoxicology as evidenced by the number of types and sizes of physical models that have been developed since then. I speculate that the type of model designed for ecotoxicological studies was dependent upon the investigators bias or past experience in either toxicology or ecology.

Ecotoxicologists with a background in toxicology developed "generic environmental models". These models contain more than one species. The species composition is selected and controlled by the investigator to include functional characteristics of some natural system such as a primary producer[20,21] or grazer.[22] Replicated model systems are initially stocked to form exactly the same species composition and numerical abundances prior to the addition of the chemical. There is no attempt to create any natural system or habitat in terms of the number and type of spatial pattern of species distributions. This bias may be derived from the primary focus toxicology has on precision and reproducibility of experimental results[23] as opposed to accuracy. If experimental units within a chemical series had unknown species of varying abundance due to field collections, then a high variance is expected. If the experiment were repeated at a later time or at a different laboratory, then different exposure-response curves could result. These potential results of high variance and difficulties with reproducibility are deemed more important to toxicologists than potential problems with accuracy of results. Therefore, efforts were made to eliminate these potential problems with variability of model systems. Experimental units have been selectively stocked with specific species. This is done to avoid the perceived precision and reproducibility problems associated with the use of intact field assemblages into laboratory systems. In addition, traditional toxicology has had monetary concerns. Testing chemicals in smaller experimental systems is less expensive than in larger ones. The resulting physical models have, therefore, been of small size, containing species of varying density and composition (sex, age, etc.) determined solely by the investigator rather than natural, intact assemblages found in the field.

Ecologists, aware of the dependency of species numbers on sample size, have had two types of responses to the use of physical models for chemical studies: either total rejection or acceptance. Investigators in the former group argue that the inability to contain large top carnivores such as lake trout, grizzly bears and whales in physical enclosures makes their application "...too re-moved from real ecosystem experiments to provide a reliable basis for manag-ing them."[24] In this view, data from whole ecosystem experiments are the only acceptable or reliable information for predicting ecosystem responses to per-turbations. This position would be impractical for generating exposure-response curves for the thousands of chemicals in natural terrestrial and aquatic environ-ments. Also, this position assumes that the "keystone species"[25] are large and known to control those processes of interest to environmental managers. Be-cause of their absence in physical models, it is argued that either no or artifactual responses to the test chemical(s) will be observed. In support of this viewpoint, Petersen and Petersen[26] showed that the direct effects of acidifica-tion of an oligotrophic lake, namely elimination of some types of fishes, resulted in increased abundances and distributions of invertebrate predators. The zooplankton populations were affected indirectly by acid rain. However, there are two important observations. First, a "keystone species" does not have to be "big". Kitchens,[27] using $1\text{-}m^2$ pieces of an eel grass community, demon-strated that grass shrimp (~2 to 3 cm in length) were responsible for controlling the species composition and densities of the meiofaunal community to approxi-mate field levels. Second, there are many important ecosystem processes that are relatively unaffected by large biota. Studies of natural watersheds in New Hampshire as large as 13 to 16 ha observed the loss of soil nutrients following the perturbation of clear cutting and addition of herbicides.[28,29] This watershed response can be observed at much smaller spatial scales. For example, Van Voris and colleagues[30] demonstrated the same loss of nutrients in small intact soil core microcosms (15 cm in diameter by 10 cm long) from an old field system following exposure to cadmium. Similar results were found by Jackson and colleagues[31] using intact cores from a forest ecosystem exposed to realistic levels of dust from a lead smelter. These studies exemplify that it is not a necessary constraint for a physical environmental model to contain *every* species to make relevant and accurate predictions for natural systems. Only those species relevant to the processes being studied need to be present. Therefore, complete rejection of physical models for ecotoxicological studies on the basis of omission of large biota is unwarranted.

There are a number of ecologists, interested in chemicals and associated ecological effects, who embrace the use of physical models containing natural assemblages of organisms. They believe that it is possible to predict the effects of chemicals in field systems despite known variability in the composition of contained species at the start of each experiment. Some of the systems devel-oped were so large, however, that a number of significant problems arose.

First, the intactness of contained communities was compromised. Oviatt and colleagues[32] collected a number of independent 0.25 m wide sediment cores and aggregated them in their replicated physical models of a coastal marine system. This method of collection was used because there was no sampling equipment available to excise intact benthic communities with a horizontal spatial scale of 1.83 m, the diameter of their cylindrical containers. This method not only resulted in a disturbed sediment sample but also changed the natural spatial relationship of the benthic species to one another within the reconstructed core.

Second, controlling or adjusting important natural forcing functions became more difficult. Menzel and Steele,[33] using plastic containers as large as 9.5 m across by 23.5 m deep, did not stir the water column to simulate water motion. The investigators knew and later demonstrated[34] that the vertical eddy-diffusivity of seawater in the plastic containers was less than that of the natural system. Subsequent studies, using these same containers,[35] injected air bubbles into the bottom of the containers to create water turbulence. This physical agitation of the water column resulted in more structurally and functionally natural phytoplankton communities in contrast to stagnant containers. However, it is important to point out that the addition of turbulence via air injection *qualitatively* corrected the unrealistic condition of water stagnancy; no attempts were made to simulate quantitatively realistic water turbulence levels in these containers. In addition, the continuous use of bubbles to create water turbulence (1) can alter the partial pressures of dissolved gases and (2) will remove organic compounds from the seawater of these systems.

Third, replication of experimental units was sacrificed presumably due to high monetary costs.[36-38] Some of these investigators have developed positions explaining why replication was not needed. For example, Gamble[39] states, "…true replication [of these systems] cannot be achieved" and additionally he predicted that "…the degree of natural variation within treated and control mesocosms is such that subtle differences between them induced by added chemicals will be extremely difficult to verify statistically." Also, Parsons[40] proposed that emphasis be placed upon reproducibility of results in identical experiments performed at different time periods rather than a single experiment with internal replication. However, such an experimental design assumes no interaction between time and the test chemical. That is, what appears to be model replication variability due to the chemical could very well be due to time or seasonal effects. The outcome of these viewpoints has been a general lack of replication for systems of this size, with a loss in statistical rigor (see Hurlbert[41] for a discussion of the statistical consequences). One of the major features associated with the use of physical models is the implicit ability to replicate them in sufficient numbers so that standard statistical techniques can be applied. A powerful quantitative tool should not be discarded because of an arbitrary choice of experimental system whose size is beyond the financial means of the project.

Fourth, as physical models have increased in size, the ability to define accurately the densities of contained biota decreased. Stephenson and colleagues[42] and Kaushik and colleagues[43] showed that the distribution of contained zooplankton became more heterogeneous with increasing container size. Both of these studies concluded that sampling strategies should be adapted for mesocosm experiments to reduce errors in estimation of zooplankton densities. These two studies successfully correlated a biotic variable (zooplankton abundance) with system size. An important question is whether the increased spatial variance of zooplankton was due to natural properties of size or the failure to simulate appropriate forcing functions. As will be discussed below (see Simu lation Conditions of Physical Models), the problem of spatial heterogeneity may be more appropriately resolved by the application of realistic simulation of water turbulence rather than a more intensive spatial and temporal sampling strategy.

Another issue of spatial scale is that of potential artifacts associated with the method of physical containment. This potential problem exists for all laboratory systems and field enclosures, especially when the model system is significantly smaller than the natural system. In aquatic model systems, container walls provide an artificial surface area greatly exceeding the area found in the natural system. Organisms can colonize and thrive on these surfaces.[44-49] The potential effects of these organisms on the chosen biotic variables, as well as the added chemical in aquatic models, has not been established. There have been two approaches to minimize wall effects: use containers as large as possible[39] (as discussed above) or apply a specific cleaning regime.[50,51] To quantify the biological effects of walls, a cleaning study should be performed using a chosen system size with and without a chemical. To my knowledge this has not been done. Such studies are needed to advance this type of modeling approach further.

There have been a number of aquatic studies that have attempted to determine the influence of container size on selected ecological variables. The causal factors responsible for the relationship between container size and ecological variables are difficult to isolate. If natural assemblages are excised, then variations in container size result in different species composition, wall effects, temperature and light profiles, etc. When conducting a size experiment, scalar variables that are controllable should be adjusted to fixed levels (e.g., water depth should be the same in all containers). These experiments would enable the investigator to isolate better the causal factors responsible for the observed size relationship. However, most studies have confounded true size effects (species composition, wall effects, etc.) with scaling variables that could have been controlled.[15,52,53] On the other hand, Perez and colleagues[54] and Kaushik and colleagues[43] varied only the cross-sectional diameter of their containers. Specifically, Perez and colleagues[54] found size-dependent differences in chemical exposure concentrations in the sediment. They concluded that increased size caused increased species richness and concomitantly higher sediment bioturbation, resulting in higher sediment exposures. Kaushik and colleagues[43] concluded that size-related adsorption of a chemical to the walls of containers resulted in the measured differences in water column exposure with concomitant inverse size-dependent effects on zooplankton. The significance of these results is that both studies demonstrated, from an exposure perspective alone, that small scale systems underestimate the ecological risks of chemicals. Although the mechanisms responsible for the size-dependent exposures differed, the implications of these results for risk assessment of chemicals using physical models were the same: spatial scale of experimental test systems significantly affects the relationship between chemical loading and associated ecological effects.

TEMPORAL SCALE

It is difficult to separate time from spatial scale when designing physical models of natural systems because temporal and spatial scale effects are intricately coupled.

When intact communities are excised for an experiment, the season or even time of day may have a significant effect on the species composition of the replicated physical models. The presence, absence, or physiological state of a particular species or community is often dependent upon the season of the year. Additionally, seasonality controls the value of major environmental variables, such as temperature, light, and rainfall.[55] These time-dependent features can influence the outcome of chemical studies.[56] Perez and colleagues[54] found in the early summer season that 10 µg/L additions of Kepone to the water column of a coastal marine model reduced the total species richness of the benthic community by a factor of five relative to controls. However, when this experiment was repeated in the late fall (Perez and colleagues), Kepone levels, which previously affected benthic communities, now had no statistically significant effect. It could be argued that the lower autumn temperature was sufficiently reduced to eliminate benthic activity. A previous study[57] exposed these same systems to an industrial plasticizer during the summer temperature maxima (~26°C) and winter minima (~1°C). The bioaccumulation rates of actively feeding bivalves were greater by an order of magnitude during the winter season than those during the summer season. These results were caused, in part, by the seasonal phytoplankton bloom, which occurred during the winter but not during the summer. This means that benthic organisms are capable of functioning actively even when temperatures are extremely low. Perhaps the absence of or insufficient numbers of, food organisms in the water column, a seasonal phenomenon, could account for the lack of Kepone toxicity during the late autumn experiment discussed above.

The time interval required for the physical model to achieve a steady state after chemical addition is another feature of temporal scale. Toxicology has grappled with this issue when interpreting differences between exposure-response curves for acute and chronic tests. Length of time is one of the critical and major differences between these two types of tests. In ecotoxicological testing, the time interval before an ecological change is detected may be very long especially at low concentrations (µg/L) of the test chemical. This is due to the mechanism by which the chemical acts on the system. Indirect effects (predator release, competition release or chemical degradation products, etc.) are not always immediately realized, especially at lower levels of chemical exposure. Similarly, some communities have different response times to a particular chemical. In aquatic systems, impacts on benthic communities usually take longer to occur than water column communities due, in part, to the longer time period required for the chemical to reach steady state in sediments.

Variable response time is also intricately coupled to size of the physical model. For example, Perez and colleagues[54] showed, in the absence of a chemical, that the timing and intensity of phytoplankton growth was dependent upon the size of the containers. Increases in container size resulted in concomitant increases in the time to achieve maximum cell densities of phytoplankton as well as the numerical value at these maxima. Therefore, we need the capability to "speed up time" and make predictions at time periods much longer than the experimental periods currently used to generate data. This capability is needed not only for predicting the time to reach "steady state" following the release of a chemical to a natural system, but also the time for recovery to occur following the application of source control technologies. The total time for recovery of a polluted coastal marine system (Narragansett Bay, RI) was estimated using physical models of various segments along a pollution gradient.[58] Direct empirical measurements of the most recalcitrant chemicals along the gradient before and after one annual cycle, without chemical additions, enabled the investigators to estimate recovery time to be over 100 years for this particular system.

SIMULATION CONDITIONS OF PHYSICAL MODELS

The construction of a physical environmental model begins by selecting and excising appropriately sized and temporally relevant component communities. For small natural systems such as "tree hole ecosystems",[59] collection of the entire system may be necessary since the biota on the "wood walls" may be critical to the dynamics of the contained water column community. For large natural ecosystems, like coastal embayments, communities developing on the intertidal and subtidal surfaces may not have to be included with water column and benthic communities because the relative size of this solid surface (viz., the "natural side walls") in the natural system does not merit its inclusion.[51] However, this assumption has never been tested experimentally. In the design of lacustrine models for acid rain studies, benthic as well as water column communities must be included, since the former subsystem significantly influences the buffering capacity of lakes.[60]

The next step involves the coupling of these component communities. This coupling may be artificial if one or more physical factors are reduced to a smaller scale. For example, Perez and colleagues[51] did not recreate the natural

8-m water column when attempting to simulate a portion of a natural coastal system. They physically "collapsed" the water column to ~1 m and coupled an intact sediment core to the water column by an air lift pump but maintained the natural ratio of sediment surface area to water column volume. The adequacy of the simulation was based upon the lack of vertical salinity and temperature stratification in the field and the assumption that exposure to the average light intensity for the 8-m water column in the laboratory would be effectively equivalent to the field system. This illustrates that it may not be necessary to recreate exactly all of the spatial features of a natural system when coupling together excised communities.

Once the component communities are appropriately coupled, it is necessary to apply relevant and realistic forcing functions. Some important forcing functions are light energy, salinity, temperature, water turnover rates, water turbulence, and rainfall. Stephenson and colleagues,[42] working with freshwater systems, found the spatial heterogeneity of zooplankton to increase as the container size increased. The problem was so acute that the authors recommended the development of sampling strategies to estimate better zooplankton abundance when models of this large size (1000 m^3 of water) are used. These studies implied that larger systems allow for more spatial heterogeneity to occur, which results in a sampling problem. However, the increased spatial heterogeneity may not have been a natural effect of size but rather the result of the relatively motionless water body in the large containers coupled with heterogenous light fields. Zooplankton migrate to and aggregate around areas of relatively high light intensities. Stephenson and colleagues[42] suggested that differences in light intensity (and a number of other factors) around the edges of the containers could have accounted for the observed spatial heterogeneity. This spatial heterogeneity of zooplankton was probably an artifact resulting from the failure to simulate realistic levels of water turbulence or water motion. Failure to simulate appropriate levels of water turbulence in large containers held at sea resulted in settling of dominant phytoplankton from the contained 30-m deep water column.[35] The point is that critical environmental forcing functions must be simulated when containerizing natural communities.

For those forcing functions dependent upon size, it may not be necessary to recreate to scale the desired physical process. As illustrated above, an 8-m water column was reduced to 1 m. Adjustment of light energy to the average light intensity of the 8-m water column in 1 m was a sufficient simulation for phytoplankton growth. Additionally, Davey and colleagues,[56] in a study of surface microlayer communities, simulated surface breaking waves by physically stirring the air–water interface with glass rods instead of using a wave tank.

Finally, after the investigator designs and develops replicated physical models, there are a myriad of assumptions, abstractions of real communities, approximate simulations, and application of selected forcing functions that may sufficiently distort the models' dynamics. As Landner and colleagues[11] have correctly asked, "…what kind of simplification is allowable if we are to use the results from the model ecosystem experiments for prediction of events in natural ecosystems?" To answer this question requires an evaluation of model performance. Such performance questions have arisen not only with physical models but also with mathematical models. The answer rests with the degree of real time divergence existing between the model and natural system being simulated in the absence of the chemical. If there is statistical equivalency between these two systems, then the exposure-response of the model can be equated to the natural system if similarly perturbed. However, if the control and natural systems diverge, then one or more of the assumptions, abstractions, simplifications, simulation conditions, etc. are incorrect or improperly realized. It is the responsibility of the investigator to establish what went wrong and how to correct it.

A correction process involves the use of a "sensitivity analysis" where some or all of the simulation conditions are varied over wide, even unrealistic, ranges to establish the qualitative and quantitative (sensitivity) response of the model system.[51] If it is found that one or more of the simulation conditions have large effects on the model system, then more accurate simulation of these conditions may be necessary to create a closer, statistical agreement between model and natural system. This process involves setting the physical forcing functions of the system to natural levels and then allowing the ecological events within the model to diverge or converge to the natural system over time. It is inappropriate to allow the biological response of the model to determine the necessary simulation conditions. For example, Levy and colleagues[61] developed physical models of the epilimnion of a reservoir. One of the simulation conditions deemed important was water turbulence. The authors compared the water turbulence levels in their natural system with their model system by measuring the rate of gypsum dissolution. Based upon these measurements, they adjusted water turbulence in their model systems using stirring propellers to match the natural system. Phytoplankton was measured in replicated systems with and without the simulation of water turbulence and in the natural reservoir system. In 30 days, statistically significant blooms occurred in the model systems with turbulence, but not in the unstirred model systems and natural reservoir system. Levy and colleagues[61] concluded that the lack of water turbulence was "...the most successful technique for ensuring good tracking and replicability." That is, a physical variable, water turbulence, adjusted to a completely unrealistic level[35,51] gave the appearance of an appropriate simulation because an uncontrolled ecological variable, phytoplankton, followed the natural system. A more plausible conclusion was that other conditions in the model system were not recreated or simulated appropriately and resulted in the phytoplankton blooms. For example, in an earlier paper, some of these same authors[62] pointed to the potential artifactual effects of walls on the dynamics of ecological events in these same model systems. They speculated that wall growth did not significantly influence plankton during the first 40 to 50 days of their experiments. However, according to Pritchard and

Bourquin,[63] "...within hours to days, a microbial film begins to develop on the walls of any container used to house natural water...." That is, organisms on the walls could have released nutrients that resulted in excessive phytoplankton growth. Simulation of water turbulence provided two conditions: uniform mixing of (1) growth stimulatory materials and (2) phytoplankton in the water column. In short, the problem was not necessarily due to an incorrect simulation of water turbulence but rather the indirect effects of wall growth coupled with sufficient physical energy in the water column to maintain phytoplankton suspension. To summarize, the designer of physical models should not compromise the realism of simulation conditions to force the ecological variables of interest to converge quantitatively to the natural system.

DISCUSSION AND RECOMMENDATIONS

Up to this point, I have examined the limitations of previous studies in an attempt to identify the critical spatial and temporal elements necessary to the successful use of physical models in ecotoxicology. Usually, natural habitats, about which we wish to make predictions, are larger than any physical model we can practically create, maintain, and simulate. Lund[64] demonstrated that even a very large (16,000 m^3) contained plankton community in the field diverged from the natural habitat; presumably perturbated responses from these models would be subject to question since the control diverged from the natural field system. Levin[7] and Turner and colleagues[65] have discussed the need to make predictions across large spatial and temporal scales both within and between habitats. If getting bigger is not going to create a better model, then how can ecotoxicological predictions be made at the spatial and temporal scales of natural systems or at least the habitat level? What follows is an attempt to outline some of the necessary requirements and considerations to set up and analyze data from physical models. The purpose is to be able to make predictions about chemical loadings at larger spatial and longer temporal scales than the physical models previously described.

SPATIAL CONSIDERATIONS

Step 1

The habitat chosen for study within an ecological system should be one that is vulnerable to chemical addition. Vulnerability is predicated upon either (1) existence of unique species, (2) trophic interactions at upper levels of the food chain, and/or (3) highest potential for exposure such as soft bottom sediments in aquatic systems where chemicals tend to accumulate at concentrations orders of magnitude higher than the overlying water column.

Step 2

Intact assemblages of organisms should be used when stocking each physical model simulating a particular habitat. Using intact assemblages assumes that every species, and its immediate environs, is equally important. Any less diverse or manipulated assemblage is arbitrary and may be insensitive or overly sensitive to the particular perturbant.

Step 3

The collection method for these intact assemblages must be nondestructive to ensure the integrity of the assemblage. In addition, the samples collected must be sufficiently large. In aquatic systems, the water column should be collected without a centrifugal pump. Zooplankton appendages and partially whole individuals were found in seawater samples following the use of a centrifugal pump (Perez unpublished). In soft bottom sediments, the horizontal scale or core diameters should not be <14.5 cm according to Blomqvist.[66] He showed that core shortening, a physical distortion, becomes more of a problem at smaller diameters. Another horizontal scale requirement is for the core to contain the largest stationary organism and its immediate environs in the chosen habitat (this is discussed below in the section "Method for Predicting Fate and Ecological Effects of Chemicals Across Spatial and Temporal Scales," Steps 1 and 2). For example, the adult parchment worm (*Chaetopterus variopedatus*) has a u-shaped tube with exit and entry openings at the sediment surface 20 to 25 cm apart. Additionally, the vertical scale should be sufficiently deep to encompass the active biotic zone of contained organisms. In lacustrine[67] and marine systems (Perez unpublished), benthic biota can penetrate down to 40 and 50 cm (however, Ates[68] cites the discovery of one tube of the tube-building anenome *Ceriantheopsis americanus* to be 180 cm long in soft bottom sediments of the Gulf of Maine).

TEMPORAL CONSIDERATIONS

Step 1

The seasonal simulation conditions required to maintain intact assemblages of organisms must be established and created for the experimental period. Critical variables (e.g., horizontal and vertical water turnover time, water turbulence, the ratio of sediment surface area to water column volume in aquatic systems, rainfall frequency and quantity, diurnal light cycles, wind related transport) have to be measured in the chosen habitat and then simulated appropriately within the experimental model so that the dynamics of the contained assemblages will continue to function as in the field.

Step 2

The residence time per unit area of mobile biota (i.e., nonstationary) should be measured within the chosen habitat. Many large (e.g., fish, bears) and small (e.g., hermit crabs, quail) mobile macrofauna are transient in time and space. If the chosen experimental period is longer than the return frequency of these organisms, then they can be added to the experimental system for the appropriate time intervals. This assumes that they can reasonably move within the confines of the physical model. This simulation is important, because it addresses one of the major criticisms leveled against physical models, namely the inability to incorporate mobile macrofauna.

METHOD FOR PREDICTING FATE AND ECOLOGICAL EFFECTS OF CHEMICALS ACROSS SPATIAL AND TEMPORAL SCALES

Objective

Establish the quantitative relationship between chemical loadings and the associated ecological effects at the spatial scale of a habitat after the chemical has reached steady state in the major compartments of the habitat.

Problem

All habitats have rare and common species. Rare individuals, such as stationary biota in sediments and soils, are distributed less frequently over the spatial domain of a habitat than common species. As a result, size of intact quadrats that contain both rare and common species may be extremely large (square meters). Collecting large undisturbed quadrats and returning them to the laboratory could be extremely difficult, if not impossible. As a result, it is probably not feasible to establish directly the spatial requirements of this objective; an indirect measuring and analysis technique is required.

Confounding the spatial problem is a temporal problem. Over time, continuous addition of a chemical results in accumulation within the various compartments of a habitat until a steady state is reached. The solid phase, sediments and soils, is the slowest compartment to reach steady state and can take years. This is due primarily to (1) relatively slow bioturbation rates by solid phase biota and (2) the depth within which biological activity occurs. Since the time frame for most ecotoxicological experiments is less than 1 year, it is not practical to measure directly ecological effects of the test chemical in sediments and soils at steady state; an indirect method is, therefore, needed.

What follows is an experimental approach outlining these issues empirically. It is directed towards aquatic environments with focus on the benthic subsystem, since a chemical added to the water column reaches steady state last and achieves the highest exposure concentration in the sediments. Conceptually, the approach applies to terrestrial environments as well.

Step 1

The size of each replicate model system will be determined by the size of the sediment compartment. Within the chosen habitat, establish the dimensions of a quadrat such that the probability of containing one individual of the largest (and presumably the rarest) stationary benthic species is high. Replicates containing at least one individual of this species are collected at random from the field site. If none of the replicated cores contains the largest organism and its immediate environs, then SCUBA divers should collect intact cores containing these organisms. Large organisms can be found by either direct observation or by their localized environmental effects (e.g., tunnels and/or entry-exit tubes at the sediment surface). One of each set of replicate cores must contain this large organism. The total number of cores needed is, therefore, predicated upon the number of (1) sampling intervals within the experiment period, (2) chemical loading rates selected, and (3) replicates within (1) and (2).

The time period of the experiment should be sufficiently long to allow for a significant quantity of the chemical (at least 20% of steady-state value) to accumulate in the surficial sediments. The experiment should simulate an annual cycle. However, this may be accomplished in a more practical manner by encompassing the warm or most biologically active time of the year, thus, one annual cycle can be approximated.

Step 2

In the laboratory, the sediment cores should be coupled to appropriately sized water columns and the necessary simulation conditions of the habitat imposed. The chosen chemical, in the form and range of initial field concentrations, in the water column should be added to each replicate model system. All water exchanges should be at these concentrations. Each level of concentration represents a specific mass or load of the chosen chemical. This step assumes (for alternative, see page 66) that water column concentrations do not exceed the water solubility of the chemical.

Step 3

The chosen functional responses, r, within each chemical loading, i (i=0,..., n levels), should be measured in the replicate systems at the fixed time intervals. At least one of the replicates sampled should contain the largest organism. After measuring the functional responses (e.g., sediment nutrient regeneration), remove these replicate systems and analyze the (1) total accumulated chemical in the sediment subsystem, C_i, and (2) species composition and associated densities within the water column and sediment subsystems.

Step 4

The time, $(t_{max})_i$, for the chemical to reach a steady state in the sediment at a given loading rate can be estimated. Assuming the chemical in sediments accumulates exponentially and reaches an asymptotic steady-state condition, it is possible to estimate both the asymptote, $(C_{max})_i$, and time to reach the asymptote (~ three time constants), $(t_{max})_i$ (Figure 1a) by regression analysis. This is the estimated time step required to reach the steady-state condition for the habitat receiving loading, i. What is needed at this point is the prediction of the structural and functional responses of the chosen ecological variables at $(t_{max})_i$.

Step 5

At each sampling interval (Step 3), estimate the densities (mean numbers per m^2) of each surviving benthic species using the replicate cores from each loading rate of chemical, i. If the largest stationary benthic organism is alive at this sampling interval, the field derived density per square meter is used for this species. Calculate the diversity index,[69] A_i using these species and density estimates. These diversity indices represent the quantitative estimate of the benthic species assemblage, A_i, at time, t, or after the accumulated chemical has achieved level, C_i. Regress the species assemblage, A_i, against time and estimate the steady-state assemblage at $(t_{max})_i$, called $(A_{max})_i$ (see Figure 1b). This species assemblage is the structural state of the benthic community when the chemical has reached a steady-state condition in the sediments under each of the chemical loading rates, i. Plotting A_{max} against each chemical loading rate, i, provides the steady-state "loading-structural response curve" (Figure 2a) for the benthic subsystem at the spatial scale of the largest stationary organism. This relationship may be extrapolated to the entire habitat if the replicate cores sample all the species whose frequency is greater than the rarest organism.

Step 6

At each sampling interval and chemical loading rate, measure replicate values for each chosen functional response (e.g., nutrient regeneration rates) of the benthic system. Additionally, one of the replicates will contain the largest stationary organism (Step 1). Excluding the replicate with the largest organism, calculate the average functional response for the remaining replicate cores at the spatial scale of the largest organism. Next add the functional response measured in the replicate with the largest organism to the response calculated above. This corrects the functional response of the hypothetical quadrat by

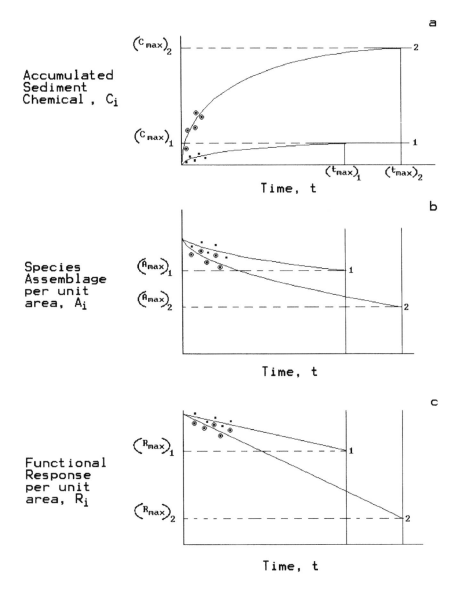

Figure 1   Time dynamics of accumulated chemical in sediments (a) and associated structural (b) and functional (c) ecological effects in two hypothetical physical models receiving chemical loadings, 1 and 2.

adding the contribution of the single large organism and other species in close proximity to this individual. The best estimate for the functional response per unit area, R_i, at this specific time interval and loading rate, i, is determined by dividing the total integrated response, just calculated, by the area of the habitat that contains one individual (or patch of individuals) of the largest stationary benthic species. Regress the functional response, R_i, against time and estimate

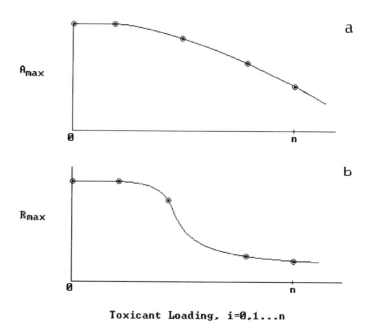

Figure 2   Physical model estimates of structural, A_{max}, and functional, R_{max}, responses of the sediment subsystem within a hypothetical habitat receiving chemical loadings, $i=0,1,...,n$.

the response at $(t_{max})_i$, called $(R_{max})_i$ (see Figure 1c). This functional response is the best estimate for the benthic community when the test chemical has reached a steady-state condition in the sediments under chemical loading rate, i. Plotting R_{max} against chemical loading rate, i, provides the steady-state "loading-functional response curve" (Figure 2b) for the benthic subsystem at the spatial scale of the largest stationary organism.

There are a number of critical points to be made about the above process. First, it was assumed that all chemical loadings would result in water column concentrations less than or equal to the water solubility of the chemical. If the concentrations exceed water solubility, the chemical cannot reach a steady-state concentration in sediments because the compound will continuously partition from the water column to sediments (and/or surface microlayer if the specific gravity of the test chemical is <1). Because these compartments will accumulate the chemical indefinitely, ecological impacts are inevitable. This assumes the chemical and/or its degradation products have some capability to elicit ecological effects at some level and/or form of exposure. For many organic chemicals contained in anthropogenic wastes, water solubilities may be exceeded in the water column at levels as low as micrograms per liter or less.

Secondly, when the water solubility is not exceeded, the success of the process outlined above is dependent upon the replicate variability both within time intervals and chemical loadings. If this variability is reasonably low to allow for convergence of data using regression models, and there is good agreement between the chosen biotic variables in the field and control systems (as outlined earlier), then this process can begin to address impacts at the habitat level of organization. Future studies need to establish whether this process is, in fact, feasible.

Thirdly, if some structural and/or functional change in the model system is found to be the result of the chemical, then it is the responsibility of the investigator to describe the consequences of these changes. For example, if a particular chemical loading results in reduced sediment resuspension by benthic communities, then the recovery time for this habitat (and entire system if this habitat has the highest accumulated load of chemicals) may be significantly increased. Therefore, when measuring the response of some ecological variable under various chemical exposures or loadings, it is essential to predict the *ecological significance* of the observed changes over the spatial and temporal scales of the habitat.

Ecotoxicology has and will continue to use physical models to make predictions about the impacts of anthropogenic materials on ecological systems. However, it has been difficult to make accurate predictions. The usual explanation for these inaccuracies is that ecosystems or habitats within ecosystems are "complex". This usually refers to the observation that multiple ecosystem experiments do not always yield consistent results. Why does this happen? Is complexity a feature that defies understanding or analysis? I would argue that variations in spatial and temporal factors, such as size-dependent species composition and seasonality, potentially change the ecotoxicological model to a degree that precludes consistent and accurate predictions of the natural system being simulated. What is needed, therefore, is to define the extent to which these spatial and temporal features can change in the models without jeopardizing the accuracy of predicted perturbation behavior. When this is done, the discipline of ecotoxicology can truly become a predictive science.

ACKNOWLEDGMENTS

Drs. Earl W. Davey, Edward H. Dettman, James F. Heltshe, and Wayne R. Munns, Jr. critically reviewed this chapter. Additionally, John A. Cardin, Roxanne L. Johnson, and R. A. Voyer provided many helpful and constructive revisions to the manuscript. Douglas A. Cory, Jr. drew and revised the figures. Two anonymous reviewers made constructive comments. Environmental Research Laboratory contribution #1435.

REFERENCES

1. Pope, A. and K. L. Goin, *High-Speed Wind Tunnel Testing,* New York: John Wiley & Sons, Inc. (1965).
2. Gilmer, T. C., *Modern Ship Design,* Annapolis, MD: U.S. Naval Institute (1970).

3. Pokrefke, T., Physical River Model Results and Prototype Response, in *Hydraulic Engineering*, S. R. Abt and J. Gessler, Eds., New York: American Society of Civil Engineers (1988), pp. 758–761.

4. Weiner, B. H., I. S. Ockene, P. H. Levine, H. F. Cuenoud, M. Fisher, B. F. Johnson, A. S. Daoud, J. Jarmolych, D. Hosmer, M. H. Johnson, A. Natale, C. Vaudreuil, and J. J. Joogasian, Inhibition of Atherosclerosis by Cod-Liver Oil in a Hyperlipidemic Swine Model, *New Engl. J. Med.* 315(14):841–846 (1986).

5. Swenberg, J. A., α_{2u}-Globulin Nephropathy: A Review of the Cellular and Molecular Mechanisms Involved and Their Implications for Human Risk Assessment, *Environ. Health Perspect.*, 101(Suppl. 6):30–44 (1993).

6. Schwartz, R. S., D. R. Holmes, and E. J. Topol, The Restenosis Paradigm Revisited: An Alternative Proposal for Cellular Mechanisms, *J. Am. Coll. Cardiol.* 20(5):1284–1293 (1992).

7. Levin, S. A., Preface, in *Ecotoxicology: Problems and Approaches*, S. A. Levin, M. A. Harwell, J. R. Kelley, and K. D. Kimball, Eds., New York: Springer-Verlag (1989), pp. v–vi.

8. Schindler, D. W., R. Wagemann, R. B. Cook, T. Rusczynski, and J. Prokopowich, Experimental Acidification of Lake 223, Experimental Lakes Area: Background Data and the First Three Years of Acidification, *Can. J. Fish. Aquat. Sci.* 37:34–354 (1980).

9. Gearing, J. N., The Role of Aquatic Microcosms in Ecotoxicological Research as Illustrated by Large Marine Systems, in *Ecotoxicology: Problems and Approaches*, S. A. Levin, M. A. Harwell, J. R. Kelly, and K. D. Kimball, Eds., New York: Springer-Verlag (1989), pp. 411–448.

10. Menzel, D. W., Summary of Experimental Results: Controlled Ecosystem Pollution Experiment, *Bull. Mar. Sci.* 27:142–145 (1977).

11. Landner, L., H. Blanck, U. Heyman, A. Lundgren, M. Notini, A. Rosemarin, and B. Sundelin, Community Testing, Microcosm and Mesocosm Experiments: Ecotoxicological Tools with High Ecological Realism, in *Chemicals in the Aquatic Environment*, L. Landner, Ed., New York: Springer-Verlag (1989), pp. 216–251.

12. Morton, R. D., T. W. Duke, J. M. Macauley, J. R. Clark, W. A. Price, S. J. Hendricks, S. L. Owsley-Montgomery, and G. R. Plaia, Impact of Drilling Fluids on Seagrasses: An Experimental Community Approach, in *Community Toxicity Testing*, STP920, J. Cairns, Jr., Ed., Philadelphia, PA: American Society for Testing and Materials (1986), pp. 199–212.

13. Grassle, J. F., R. Elmgren, and J. P. Grassle, Response of Benthic Communities in MERL Experimental Ecosystems to Low Level Chronic Additions of No. 2 Fuel Oil, *Mar. Environ. Res.* 4:279–297 (1981).

14. Takahashi, M., W. H. Thomas, D. L. R. Seibert, J. Beers, P. Koeller, and T. R. Parsons, The Replication of Biological Events in Enclosed Water Columns, *Arch. Hydrobiol.* 76(1):5–23 (1975).

15. Kuiper, J., U. H. Brockmann, H. van het Groenewoud, G. Hoornsman, and K. D. Hammer, Influences of Bag Dimensions on the Development of Enclosed Plankton Communities during POSER, *Mar. Ecol. Prog. Ser.* 14:9–17 (1983).

16. Pilson, M. E. Q., C. A. Oviatt, G. A. Vargo, and S. L. Vargo, Replicability of *MERL* Microcosms: Initial Observations, in *Advances in Marine Environmental Research*, EPA-606/9–79–035, F. S. Jakoff, Ed., Narragansett, RI: U.S. EPA Environmental Research Laboratory (1979).

17. Steele, J. H. and E. W. Henderson, Simulation of Vertical Structure in a Planktonic Ecosystem, *Scott. Fish. Res. Rep.* 5:1–27 (1976).

18. Winer, B. J., *Statistical Principles in Experimental Design*, New York: McGraw-Hill (1971).

19. Gleason, H. A., On the Relation Between Species and Area, *Ecology* 3:158–162 (1922).

20. Gillett, J. W. and J. D. Gile, Pesticide Fate in Terrestrial Laboratory Ecosystems, *Int. J. Stud.* 10:15–22 (1976).

21. Taub, F. B., A Biological Model of a Freshwater Community: A Gnotobiotic Ecosystem, *Limno. Oceanogr.* 14:136–142 (1969).

22. Metcalf, R. L., I. P. Kapoor, P.-Y. Lu, C. K. Schuth, and P. Sherman, Model Ecosystem Studies of the Environmental Fate of Six Organochlorine Pesticides, *Environ. Health Perspect.* 4:35–44 (1973).

23. Giesy, J. P. and P. M. Allred, Replicability of Aquatic Multispecies Test Systems, in *Multispecies Toxicity Testing*, J. Cairns, Jr., Ed., New York: Pergamon Press (1985), pp. 187–247.

24. Schindler, D. W., Detecting Ecosystem Responses to Anthropogenic Stress, *Can. J. Fish. Aquat. Sci.* 44(Suppl. 1):6–25 (1987).

25. Paine. R. T., Food Web Complexity and Species Diversity, *Am. Nat.* 100:65–75 (1966).

26. Petersen, R. C., Jr. and L. B.-M. Petersen, Ecological Concepts Important for the Interpretation of Effects of Chemicals on Aquatic Systems, in *Chemicals in the Aquatic Environment*, L. Landner, Ed., New York: Springer-Verlag (1989), pp. 165–192.

27. Kitchens, W. M., Development of a Salt Marsh Microecosystem, *Int. J. Environ. Stud.* 13:109–118 (1979).

28. Likens, G. E., F. H. Bormann, N. M. Johnson, D. W. Fisher, and R. S. Pierce, Effects of Forest Cutting and Herbicide Treatment on Nutrient Budgets in the Hubbard Brook Watershed Ecosystem, *Ecol. Monogr.* 40:23–47 (1970).

29. Bormann, F. H., G. R. Likens, T. G. Siccama, R. S. Pierce, and J. S. Eaton, The Export of Nutrients and Recovery of Stable Conditions Following Deforestation at Hubbard Brook, *Ecol. Monogr.* 44:255–277 (1974).

30. Van Voris, P., R. V. O'Neill, W. R. Emanuel, and H. H. Shugart, Functional Complexity and Ecosystem Stability, *Ecology* 61(6):1352–1360 (1980).

31. Jackson, D. R., W. J. Selvidge, and B. S. Ausmus, Behavior of the Heavy Metals in the Forest Microcosms: Effects on Nutrient Cycling Processes, *Water Air Soil Pollut.* 10:13–18 (1978).

32. Oviatt, C. A., M. E. Q. Pilson, S. W. Nixon, J. B. Frithsen, D. T. Rudnick, J. R. Kelly, J. F. Grassle, and J. P. Grassle, Recovery of a Polluted Estuarine System: A Mesocosm Experiment, *Mar. Ecol. Prog. Ser.* 16:203–217 (1984).

33. Menzel, D. W. and J. H. Steele, The Application of Plastic Enclosures to the Study of Pelagic Marine Biota, *Rapp. P. v. Reun. Cons. Int. Explor. Mer.* 173:7–12 (1978).

34. Steele, J. H., D. M. Farmer, and E. W. Henderson, Circulation and Temperature Structure in Large Marine Enclosures, *J. Fish. Res. Board Can.* 34:1095–1104 (1977).

35. Eppley, R. W., P. Koeller, and G. T. Wallace, Stirring Influences the Phytoplankton Species Composition within Enclosed Columns of Coastal Water, *J. Exp. Mar. Biol. Ecol.* 32:219–239 (1978).

36. Grice, G. D., M. R. Reeve, P. Koeller, and D. W. Menzel, The Use of Large Volume, Transparent, Enclosed Sea-surface Water Columns in the Study of Stress on Plankton Ecosystems, *Helgol. Wiss. Meeresunters.* 30:118–133 (1977).

37. Kuiper, J., Ecotoxicological Experiments with Marine Plankton Communities in Plastic Bags, in *Marine Mesocosms*, G. D. Grice and M. R. Reeve, Eds., New York: Springer-Verlag (1982), pp. 181–193.

38. Reeve, M. R., J. C. Gamble, and M. A. Walter, Experimental Observations on the Effects of Copper on Copepods and Other Zooplankton: Controlled Ecosystem Pollution Experiment, *Bull. Mar. Sci.* 27(1):92–104 (1977).

39. Gamble, J. C., Mesocosms: Statistical and Experimental Design Considerations, in *Coastal and Estuarine Studies*, C. M. Lalli, Ed., New York: Springer-Verlag (1990), pp. 188–196.

40. Parsons, T. E., The Future of Controlled Ecosystem Enclosure Experiments, in *Marine Mesocosms*, G. D. Grice and M. R. Reeve, Eds., New York: Springer-Verlag (1982), pp. 411–418.

41. Hurlbert, S., Pseudoreplication and the Design of Ecological Fish Experiments, *Ecol. Monogr.* 54:187–211 (1984).

42. Stephenson, G. L., P. Hamilton, N. K. Kaushik, J. B. Robinson, and K. R. Solomon, Spatial Distribution of Plankton in Enclosures of Three Sizes, *Can. J. Fish. Aquat. Sci.* 41:1048–1054 (1984).

43. Kaushik, N. K., K. R. Solomon, G. L. Stephenson, and K. E. Day, Use of Limnocorrals in Evaluating the Effects of Pesticides on Zooplankton Communities, in *Community Toxicity Testing*, STP920, J. Cairns, Jr., Ed., Philadelphia, PA: American Society for Testing and Materials (1986), pp. 269–290.

44. Characklis, W. G., Attached Microbial Growths: Attachments and Growth, *Water Res.* 7:1113–1127 (1973).

45. Adams, S. M., *Evaluation and Critique of in situ Experimental Systems for Investigating Effects of Stress in Lentic Ecosystems*, Environ. Sci. Div. Publ. No. 1949, Contract Number W-7405-eng-26, TM-8270, Oak Ridge, TN: Oak Ridge National Laboratory (1982).

46. Giesy, J. P. and E. P. Odum, Microcosmology: Introductory Comments, in *Microcosms in Ecological Research*, J. P. Geisy, Ed., Springfield, VA: Technical Information Center (1980), p. 1.

47. Strickland, D. H., Between Beakers and Bays, *New Sci.* 33:276–278 (1967).

48. Howell, J. A., C. T. Chi, and U. Pawlowsky, Effect of Wall Growth on Scale-Up Problems and Dynamics Operating Characteristics of the Biological Reactor, *Biotechnol. Bioeng.* 14:253–265 (1972).

49. Corpe, W. A., Periphytic Marine Bacteria and the Formation of Microbial Films on Solid Surfaces, in *Ocean Environment Microbial Activities*, R. R. Colwell, and R. Y. Morita, Eds., Baltimore, MD: University Park Press (1972), pp. 411–415.

50. Harte, J., D. Levy, J. Rees, and E. Saegebarth, Making Microcosms an Effective Assessment Tool, in *Microcosms in Ecological Research*, J. P. Geisy, Ed., Springfield, VA: Technical Information Center (1980), pp. 105–137.

51. Perez, K. T., G. E. Morrison, N. F. Lackie, C. A. Oviatt, S. W. Nixon, B. A. Buckley, and J. F. Heltshe, The Importance of Physical and Biotic Scaling to the Experimental Simulation of a Coastal Marine Ecosystem, *Helgol. Wiss. Meeresunters.* 30:144–162 (1977).

52. Adler, D., M. Amdurer, and P. H. Santschi, Metal Tracers in Two Marine Microcosms: Sensitivity to Scale and Configuration, in *Microcosms in Ecological Research*, J. P. Geisy, Ed., Springfield, VA: Technical Information Center (1980), pp. 348–368.

53. Giddings, J. M. and G. K. Eddlemon, The Effects of Microcosm Size and Substrate Type on Aquatic Microcosm Behavior and Arsenic Transport, *Arch. Environ. Contam. Toxicol.* 6:491–505 (1977).

54. Perez, K. T., G. E. Morrison, E. W. Davey, N. F. Lackie, A. E. Soper, R. J. Blasco, D. L. Winslow, R. L. Johnson, P. G. Murphy, and J. F. Heltshe, Influence of Size on Fate and Ecological Effects of Kepone in Physical Models, *Ecol. Appl.* 1(3):237–248 (1991).

55. Leppakoski, E. and E. Bonsdorff, Ecosystem Variability and Gradients: Examples from the Baltic Sea as a Background for Hazard Assessment, in *Chemicals in the Aquatic Environment*, L. Landner, Ed., New York: Springer-Verlag (1989), pp. 22–58.

56. Davey, E. W., K. T. Perez, A. E. Soper, N. F. Lackie, G. E. Morrison, R. L. Johnson, and J. F. Heltshe, Significance of the Surface Microlayer to the Environmental Fate of di(2-ethylhexyl)phthalate Predicted from Marine Microcosms, *Mar. Chem.* 31:231–269 (1990).

57. Perez, K. T., E. W. Davey, N. F. Lackie, G. E. Morrison, P. G. Murphy, A. E. Soper, and D. L. Winslow, Environmental Assessment of a Phthalate Ester, di(2-ethylhexyl)phthalate (DEHP), Derived from a Marine Microcosm, in *Aquatic Toxicology and Hazard Assessment*, STP802, W. E. Bishop, R. D. Caldwell, and B. B. Heidolph, Eds., Philadelphia, PA: American Society for Testing and Materials (1983), pp. 180–190.

58. Perez, K. T., E. W. Davey, J. Heltshe, J. A. Cardin, N. F. Lackie, R. L. Johnson, R. J. Blasco, A. E. Soper, and E. Read, *Recovery of Narragansett Bay, RI: A Feasibility Study*, Contribution No. 1148, Narragansett, RI: U.S. EPA, Office of Research and Development (1990).

59. Walker, E. D., D. L. Lawson, R. W. Merritt, W. T. Morgan, and M. J. Klug, Nutrient Dynamics, Bacterial Populations, and Mosquito Productivity in Tree Hole Ecosystems and Microcosms, *Ecology* 72(5):1529–1546 (1991).

60. Schiff, S. L. and R. F., Anderson. Limnocorral Studies of Chemical and Biological Acid Neutralization in Two Freshwater Lakes, *Can. J. Fish. Aquat. Sci.* 4(Suppl. 1):173–187 (1987).

61. Levy, D., G. Lockett, J. Oldfather, J. Rees, E. Saegebarth, R. Schneider, and J. Harte, Realism and Replicability of Lentic Freshwater Microcosms, in *Validation and Predictability of Laboratory Methods for Assessing the Fate and Effects of Contaminants in Aquatic Ecosystems*, STP865, T. P. Boyle, Ed., Philadelphia, PA: American Society for Testing and Materials (1983) pp. 43–56.

62. Dudzik, M., J. Harte, A. Jassby, E. Lapan, D. Levy, and J. Rees, Some Considerations in the Design of Aquatic Microcosms for Plankton Studies, *Int. J. Environ. Stud.* 13:125–130 (1979).

63. Pritchard, P. H. and A. W., Bourquin, In *Advances in Microbial Ecology, Vol. 7*, K. C. Marshall, Ed., New York: Plenum Press (1984), pp. 133–205.

64. Lund, J. W. G., Preliminary Observations on the Use of Large Experimental Tubes in Lakes, *Verh. Int. Verein. Theor. Angew. Limnol.* 18:71–77 (1972).

65. Turner, M. G., V. H. Dale, and R. H. Gardner, Predicting Across Scales: Theory Development and Testing, *Landsc. Ecol.* 3(3/4):245–252 (1989).

66. Blomqvist, S. Reliability of Core Sampling of Soft Bottom Sediment—an *in situ* Study, *Sedimentology* 32:605–612 (1985).

67. Strayer, D., The Benthic Micrometazoans of Mirror Lake, New Hampshire, *Arch. Hydrobiol./Suppl.* 72(3):287–426 (1985).

68. Ates, R. M. L., Fishes That Eat Sea Anemones, a Review, *J. Nat. Hist.* 23:71–79 (1989).

69. Pielou, E. C., The Measurement of Diversity in Different Types of Biological Collections, *J. Theoret. Biol.* 13:131–144 (1966).

CHAPTER 5

Design and Analysis of Multispecies Experiments

Eric P. Smith

INTRODUCTION

Multispecies experiments vary in the way they are physically designed, in the way they are constructed, in the cost to carry out the study, and in the objectives of the study. Stream mesocosms cannot be expected to be similar in construction to pond mesocosms nor can large pond studies be expected to be similar to "tub" mesocosms. A multispecies study may be designed to negate the laboratory toxicity of a pesticide or to estimate the concentration that affects half of the organisms. Although considerably different in terms of the physical systems, and sometimes objectives, multispecies systems share some similarities. Typically, these studies involve a single stressor. Also, the analysis of multispecies data involves either a regression or analysis of variance (ANOVA) approach for interpreting results. While it is possible to list many objectives, generally the purpose of these studies is to investigate impacts of suspected pollutants and to use these studies as a bridge between laboratory studies and field sampling of ecosystems. Because of these similarities, it is possible to talk about the general problem of design and analysis of multispecies tests.

This chapter deals with the important aspect of the design and interpretation of multispecies studies. Our primary focus is the use of statistics as an aid in designing and analyzing these experiments. Multispecies experiments are viewed as experiments that can be manipulated to answer better the questions of interest regarding effects of stressors. There are four important aspects to the design and use of multispecies tests. First is the construction (physical design) of the test systems. This area is outside our interest as the construction involves ecological issues (Dudzik et al. 1979). However, an important element of construction is the physical scale of the test system. The scale of the system determines the types of hypotheses that may be addressed and the endpoints that are used to assess stressor effects. The question of scale is addressed in another chapter in this text.

The second aspect of multispecies tests is the design of the experiment. As most multispecies tests involve a single stressor, only that case is dealt with. Topics in the design of these tests include the selection of the doses of the stressor (how many and the spacing of doses), the number of replicates for each dose, and the allocation of the doses to the test systems. Also part of the design phase of the experiment is the determination of objectives and hypotheses. Hypotheses and objectives are useful for identifying models or types of models that may be used to determine statistical objectives and criteria. With these criteria, statistical measures may be used to compare different designs and either select reasonable designs or eliminate some designs.

Multispecies tests have been criticized as having low replicability and reproducibility. Replicability refers to the ability of two test units that receive the same dose to produce the same results and is generally associated with "within treatment" variation (Giesey and Allred 1985). Reproducibility refers to the ability of the test system to provide similar results when used by different researchers at different locations. The ability of the design to have high replicability and reproducibility depends on the complexity of the system, the endpoints that are measured, and the laboratory conditions. To achieve high replicability and reproducibility requires standardization of the test procedures and care must be taken to insure high quality control of the results.

The final statistical component of multispecies tests is the analysis of test results. There are many approaches to the analysis of data resulting from these tests, and the choice of method depends on the objectives of the study. Some general approaches to the analysis of the data are given without much detail. Also, distinctions and similarities between regression and ANOVA approaches are described.

DESIGN AND DESIGN ISSUES

Multispecies studies are not simple experiments. Researchers do not want them to be simple. The rationale behind using a more realistic experimental approach is that laboratory toxicity studies are too simple, and the results may not be applicable to ecosystems. The desire for complexity leads to multiple hypotheses of interest, an interest in both testing and estimation, and frequent limitations in the number of replicate systems that may be included in a study. In this section, we will not show that statistical design will solve all the problems associated with multispecies test studies. Rather, we focus on how statistics can be used to aid in developing a sensible experiment and how an

Table 1   Some Objectives in Multispecies Studies

- Detect an effect or confirm an effect (not detected in a toxicity test)
- Indicate no effect (negate a toxicity test)
- Estimate a "safe" concentration
- Estimate an endpoint of interest (LC_{50}, etc.)
- Estimate the time required for an event to occur (recovery of system)
- Describe the relationship between the chemical and the effects of the chemical

experimenter can control the ability of the experiment to answer the main questions and address the main objectives of interest.

Although there is not a typical multispecies study, many share similarities from a statistical view. The studies are typically limited in terms of the number of test units (ponds, artificial streams, etc.) that may be used as the cost of construction may be high. For example, Liber et al. (1992) describe an experiment in which the effects of a pesticide on zooplankton are studied. Three blocks of enclosures were placed along the shoreline. Each block consisted of four enclosures; one acted as a control, two received treatments (concentrations of a pesticide), and the fourth was used for a second study. Zooplankton were collected over time and identified at the species level.

Multispecies tests often share a common set of objectives. Some of these are presented in Table 1. Although this is not an exclusive list, most tests focus on one or several of these objectives. The objectives are also valuable for statistical assessment of design as they lead to criteria for determining optimal design.

In the design of a multispecies test system, the researcher has control over several factors that influence the ability to meet the above objectives. These include the number of replicates, the treatment levels, the allocation of replication as well as control over the quality of the experiment. By a replicate, we mean an additional test unit. The test unit is the unit that receives the treatment. A test unit is differentiated from a measurement unit in that the test unit is the smallest unit that can be treated, provided that two such units can receive different treatments. Thus, two measurements on one mesocosm are viewed as information on one unit. Measurement units refer to the unit that measurements are made on. When multiple measurements are made on the same test unit these are sometimes referred to as subsamples and reflect variation within a test unit. The replicate or test unit is usually the mesocosm, pond, artificial stream, etc. to which the treatments are applied and reflect variation between test units (the more important variation). The treatment level is usually thought of as a dose or concentration. However, in some cases, it may refer to a level of a stressor and represent exposure to a stressor. Here, we use treatment level and dose interchangeably.

In addition to the above control factors, the researcher has some control of some of the external variability by using statistical blocks. Latin Square designs, which use two sources of blocking, and randomized block designs (one blocking component) are especially useful for field experiments where there may be naturally occurring gradients (wind, moisture, etc.). Unfortunately, in many multispecies testing situations, control of variation through blocking is expensive as the number of test units is constrained by cost and precious degrees of freedom are used up (but see, for example, van Voris et al. 1985).

Accurate determination of the number of replicates to use in a multispecies test is a difficult task due to the multiobjective nature of the study. Sample size must be sufficiently large to enable estimation of the safe dose, no observable effect level (NOEL) and related parameters, provide adequate power for testing hypotheses about differences in concentration effects, and provide information about the dose-response curve. Factors that affect the sample size are the number of dose levels, the range of the doses, and the magnitude of natural variability. In many studies, inferences are limited due to the small number of multispecies test units that are used.

When the cost of the test system is not restrictive, it is possible to develop statistically sound designs. Designs can be constructed that have high power for detecting differences and can also estimate model parameters and related quantities effectively.

When the cost of the test system is restrictive, greater care is needed in the design phase. It is advisable to have as much information as possible available to help choose the concentrations of the chemical, and prior information is useful for setting the number of replications. When the total number of multispecies test systems is fixed due to cost constraints, the power of the statistical tests and other criteria can be improved by proper design selection. The factors that contribute to the power are the range of the dose levels, the error variance, and unknown parameters. A useful way to approach the experiment is through a response surface model.

There is great diversity in the multispecies test literature concerning the location and number of dose levels and the number of replications made at each level (Crow and Taub 1979, Giddings and Eddelmon 1979, Isensee 1976, Touart, 1988). The United States Environmental Protection Agency (Touart 1988), for example, favors four concentrations and three replications at each concentration for evaluation of pesticide effects. Equal replication of concentrations cannot lead to designs that are optimal for all criteria. When the focus is on concentrations that are near safe levels, replications with considerably higher concentrations may not provide much statistical information.

The design of an experiment often starts (or at least should start) with objectives and a statistical model. Recent interest in the statistical design of multispecies studies has focused on the difference between models based on an ANOVA approach vs. a regression approach (Liber et al. 1992, Stephan and Rogers 1985, Graney et al. 1989). In the ANOVA approach, one starts by choosing the number of dose levels and the number of test units. Usually, the design is balanced so the number of replicates is the number of units divided by the number of dose levels. This gives rise to a study with, for example, m replicates and k treatments. Designs with an uneven number of replicates per treatment level are possible but have not been used often. The final step is to choose the levels of the treatment (concentrations or doses of the chemical). The statistical design of the ANOVA experiment is typically discussed in terms of a hypothesis. For the simple case of a single chemical, this becomes simply the hypothesis of no difference between means. Then the design can be assessed in terms of how likely it is that the hypothesis is rejected when there are differences between the means. In cases where the number of replicates is not severely limited, the amount of difference between means and the variation in test units can be used to estimate the number of test units that are required.

In the regression view of design, the number of treatment levels is the same as the number of observations. The test units are assigned a treatment level in a random fashion (Liber et al. 1992). Then a regression model is fit to the data and analyzed. The number of levels could be chosen in a similar manner to the ANOVA approach, but, instead of differences in means, the focus is on the parameters of the regression model.

ANOVA VS. REGRESSION

In comparing the approaches, some misconceptions have arisen that need to be addressed. The main misconception deals with differences between the approaches. From an analysis view there is really no difference between the two approaches. This misconception about a difference has arisen because analysis of variance programs tends to produce different output than regression programs. In fact, the two methods for analyzing data can produce equivalent analyses depending on how error terms are calculated. In a usual ANOVA approach to analyzing data from a designed experiment with one continuous factor, it is not uncommon to summarize results in terms of an overall test with polynomial contrasts (Kleinbaum et al. 1988, p. 374). An example is given in Table 2. If the same data are analyzed by a regression model, the results are summarized in Table 3.

If the simple linear regression model is fit, the results are given in Table 3a, and the analysis appears to be different. However, if a cubic model is fit to the

Table 2   Example of an ANOVA Table Decomposed into Orthogonal Polynomials for a Mesocosm Study With 12 Units and 4 Dose Levels

Source	Sum of squares	Degrees of freedom	Mean square	F
Doses	17.54	3	5.85	2.87
Linear	12.42	1	12.42	6.08
Quadratic	4.61	1	4.61	2.26
Cubic	0.51	1	0.51	0.25
Error	16.32	8	2.04	
Total	33.86	11		

Table 3   Example of a Regression Analysis Table for the Same Data as in Table 2

a. Analysis table based on a simple linear regression; x represents the dose

Source	Sum of squares	Degrees of freedom	Mean square	F
Model	12.42	1	12.42	5.79
x	12.42	1	12.42	
Error	21.44	10	2.14	
Total	33.86	11		

b. Analysis based on a cubic model with sequential sum of squares

Source	Sum of squares	Degrees of freedom	Mean square	F
Model	17.54	3	5.85	2.87
x	12.42	1	12.42	6.08
x^2 adjusted for x	4.61	1	4.61	2.26
x^3 adjusted for x^2 and x	0.51	1	0.51	0.25
Error	16.32	8	2.04	
Total	33.86	11		

data, one obtains Table 3b, and the results are equivalent with those in Table 2. With the ANOVA approach, the decomposition into (orthogonal) polynomials leads to the same results as the sequential sum of squares regression approach (sequential sum of squares represents the amount the sum of squares for the model changes when a term is added). As the model changes from a linear regression model to a quadratic model, the model sum of squares increases by 4.61. Thus, from an analysis point of view, the two methods produce equivalent results.

The misconception about the analyses has led to a belief that the ANOVA approach is not as powerful as the regression approach. This may be true depending on what is compared to what. For example, suppose that the same design is analyzed with regression and ANOVA techniques. If the relationship between dose and the multispecies endpoint is linear, and a linear regression is fit, then the regression method has more degrees of freedom for error and fewer for the model (Table 3a) and thus is more likely to indicate an effect than the ANOVA model. However, if one uses the ANOVA approach and fits a regression model unconditional on the outcome of the ANOVA test of equality of means and pools the higher order polynomials into the error, then both methods are equally likely to indicate a treatment effect (compare Table 3b and 2).

If the experiment differs in design, then it is difficult to say which testing approach is better. If the true relationship between the multispecies endpoint and the chemical concentration is linear, and a linear regression is fit, then the regression test will generally be a better test to indicate an effect. However, if the incorrect model is fit, then the ANOVA model may be better.

What then are the differences between the approaches? One difference can be seen by considering Table 3a. Here, a model has been selected. We note that, in Table 2, both the quadratic and cubic terms are not significant (at the 5%

level). Thus, another view is that the model is "reduced" to a linear regression, and the ANOVA table is given in Table 3a. What has been done is to add the sum of squares for the cubic and quadratic terms to the error term. In the ANOVA model of Table 2, the error term represents a "pure" error term that is model-independent and is interpreted as variation between replicates receiving the same concentration of chemical. In the regression model of Table 3b, the interpretation is the same. However, for the model in Table 3a, the view is that the error represents the error about the model and represents variation among replications plus errors in the model. Only if the model is correct (which is usually assumed) can this be viewed as replication variation. If there are replicates of each concentration, then the ANOVA approach is equivalent to the "lack-of-fit" approach to regression analysis (Kleinbaum et al. 1988, p. 237). In applications with small sample sizes, it is advantageous to add the nonsignificant polynomial terms to the error term. This may give the illusion that the ANOVA approach is less powerful than the regression approach.

A second difference is more philosophical. When statisticians think about analysis of variance, the focus is usually on testing and, in the case described above, on testing for differences due to concentrations of the chemical. In regression analysis, the focus may be on testing. However, it may also be on prediction of new observations, variance of coefficients, or on predicting safe concentrations or other summary measures. Thus, there is not so much a difference in design as there is in analysis. The ANOVA approach is associated with the "hypothesis testing" view (Stephan and Rogers 1985), which focuses on comparison of means. The regression focus is on models and the modeling process. This difference in focus leads to a different design criteria. Thus, it seems that the regression view offers a more general approach to the design of multispecies tests, especially if the number of replicates is small. Hence, from a statistical view, the issue is not between analysis of variance and regression. These terms refer to models. In terms of designs, the design used in Tables 2 and 3, regardless of the number of replicates, is a completely randomized design. The real issues of interest are how the data are to be analyzed, with what model one is to work, and how to design the study to meet objectives related to the model and analysis.

DESIGN EVALUATION

How can statistical theory and the above discussion aid in the design of multispecies tests? There is an abundance of theoretical work and practical guidance for designing experiments (Cochran and Cox 1957, Myers 1976). Most of the work on bioassays (Finney 1978) is of use. The work in bioassays focuses primarily on the problem of choosing levels that optimize the criteria of estimating the LC_{50}. The idea is to pick a criterion, such as the estimation of the LC_{50}, then decide how to allocate the doses and decide how many doses are needed to optimize the criteria. Table 4 presents some of the possible

Table 4   Possible Statistical Design Criteria

- Power of the test of the hypothesis of no effect (ANOVA model)
- Power of the test of no linear effect (regression model)
- Power of the test of no relationship (regression model)
- Estimation of parameters in regression model — variance of coefficients
- Estimation of LC_α (regression model)
- Estimation of NOEC (ANOVA model)
- Prediction of response
- Prediction of time response occurs

statistical criteria that are of interest in multispecies dose response studies. Some criteria and their effect on design are described below.

When the emphasis is on testing, the sensitivity of the study is constrained by what are called noncentrality parameters. In regression analysis, this is a function of the size of the regression parameters (power increases with increasing size), the variance (power decreasing with increasing variance), and the choice of the concentrations. One way to compare different designs statistically is to compare the theoretical ability of designs to detect differences when differences exist. Designs with replication can be used in many ways. First, means may be compared at different doses. Alternately, one can test for a slope different from zero in a simple linear regression model. One may also use as a criterion the test comparing the control (zero dose) with other doses or use a criterion based on polynomial models or nonlinear models. Here, the first two criteria are used to compare some models. The first criterion is based on the test of no differences in the means of the doses. The noncentrality parameter for the test comparing means is given by

$$NCP_{ANOVA} = \frac{\sum_{i=1}^{k} n_i(\mu_i - \mu)^2}{k\sigma^2}$$

where k is the number of doses, n_i the number of replicates for dose i, μ_i the mean of the ith dose, and μ the overall mean.

The test for the significance of a simple linear regression model uses a statistic based on the estimated slope. The noncentrality parameter for this test is given by

$$NCP_{linear} = \frac{\beta^2 \sum_{i=1}^{n} (x_i - \bar{x})^2}{\sigma^2}$$

where β is the true slope, n is the total number of replicates, x_i is the dose, and σ^2 is the true variance. For a theoretical approach to produce reasonable estimates of power, the slope parameter and the variance need to be known to a reasonable level. However, different designs with the same variance and slope can be compared. In this case, only the dose levels and number of observations are relevant. Fixing the slope and variance, we can now address several questions. How should doses be spaced to achieve the highest power? How much power is gained by using a regression analysis instead of an ANOVA model? Is an unbalanced design worse than a balanced design? These questions can be addressed by starting with a statistical model and considering different designs for estimating and testing that model.

To compare the designs, the doses were set to values between 0 and 100, the variance was set at 2500, and the slope set at -1 (the simple linear regression model was y = 100 – x). These values were chosen to give a good range on the power calculations. Power was calculated for both a simple linear regression model and an analysis of variance model. The theoretical power for eleven different designs is given in Table 5. First, note that the power is the same for the regression and ANOVA approach to testing for the first design only (as there are only two doses). Second, the greatest power is achieved when the doses are as far apart as possible (D1) and as the separation between doses decreases the power decreases. As the number of doses increases, the power decreases. The decrease in power is more dramatic for the ANOVA test than the regression test. Designs that have more replication in the center (D10 or D7) have less power than similar designs that have a more even distribution of replication (D4 or D3) or designs that have more replicates at the extremes (compare D9 and D8).

If we know that the model is linear, then the best design (D1) is one that has only two concentrations, and these are spread out as far as possible and half of the replicates placed at each concentration (Smith and Mercante 1989, Finney 1978). The superiority of the design with two levels on many criteria can be seen in Table 5 and Figure 1. Besides power, useful criteria are the variance of a predicted value at the dose x_0

$$\text{Var}[\hat{y}(x_0)] = \sigma^2[1 + \frac{1}{n} + \frac{(x_0 - \bar{x})^2}{S_{xx}}]$$

or the variance of the mean prediction at x_0

$$\text{Var}[\hat{\bar{y}}(x_0)] = \sigma^2[\frac{1}{n} + \frac{(x_0 - \bar{x})^2}{S_{xx}}]$$

where n is the total number of observations (12 in Table 5), σ^2 is the variance about the regression line, x is the mean of the x values, and S_{xx} is the sum of squares for x. The above formulae indicate that the variance in prediction is a function of the variance in the data and the doses (x values). If the variance of the data (σ^2) is fixed, then the prediction variance is made small by spreading

Table 5   Comparison of Different Designs for a Single Toxicant Using 12 Replicates; Dose Levels are Expressed as a Percentage of the Maximum Dose

Design	Dose levels	Sample Allocation	Percent PVi>PV1[a]	Percent increase APV[b]	Percent increase max PV[c]	Power Regression	Power ANOVA[d]
D1	0,100	1:1	—	—	—	0.88	0.88
D2	0,50,100	1:1:1	100	13	25	0.72	0.56
D3	0,33,66,100	1:1:1:1	100	20	40	0.64	0.37
D4	0,20,40,60,80,100	1:1:1:1:1:1	100	30	57	0.57	0.20
D5	regression	1	100	38	77	0.49	d
D6	0,50,100	1:2:1	99	25	50	0.60	0.44
D7	0,2,40,100	1:2:2:1	61	47	141	0.60	0.34
D8	0,33,66,100	1:2:2:1	99	37	73	0.51	0.28
D9	0,33,66,100	2:2:1:1	68	32	98	0.60	0.34
D10	0,20,40,60,80,100	1:2:3:3:2:1	100	57	113	0.41	0.14
D11	0,100	9:3	63	33	100	0.77	0.77

[a] PVi is the variance of a predicted observation for design i

[b] APV is the average variance of predicted observations

[c] max PV is the largest predicted variance for the design

[d] no ANOVA test is possible as there are no degrees of freedom for error

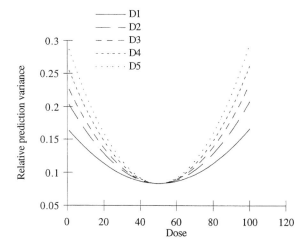

Figure 1 Comparison of relative prediction variance for several designs with different spacings of doses. Designs are described in Table 5.

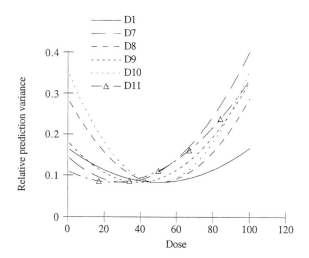

Figure 2 Comparison of relative prediction variance for several designs with different spacings or allocation of replication. Designs are described in Table 5.

the doses apart to make S_{xx} large. Hence D1 is optimal. Table 5 and Figures 1 and 2 consider the variance of the mean prediction, and, in the figures, the variance is omitted in the calculations (hence we refer to it as the relative prediction variance). The prediction variance was compared over the whole dose region using three criteria: how often the first design had the lowest variance, how much of an increase there was on the average when another design was used, and what was the largest increase. Some of the variances are plotted in Figures 1 and 2 for different designs.

Figure 1 displays the variance curve for designs that are balanced and evenly spaced but differ in the number of doses. As the number of doses increases, so does the variance. The worst case is for the case D5, which has only one observation per dose. Note also in the display that the variance is lowest in the center of the design. This is one reason why emphasis in bioassays is on estimation of the LC_{50}. Figure 2 displays prediction variance for designs that are unbalanced. Again, the design with two doses is generally superior to other designs, especially those that concentrate doses towards the center. Some improvement may be achieved if interest is in estimation of low dose effects. If more replication is placed toward one end (D11) or lower doses are used (D7), then the prediction variance is decreased for the lower doses. Note that the penalty paid is that the variance is considerably larger for greater doses with these designs.

Yet another approach to design is based on estimation of endpoints such as LC_{50} values. The problem of estimation of these endpoints and confidence limits from a regression model is referred to as the calibration problem. The solution is given in Finney (1978) or Liber et al. (1992). The estimate of the dose value x that produces a response y using a simple linear regression is given by $(y-b_0)/b_1$ where b_0 and b_1 are the estimated regression coefficients. Upper and lower limits are given by

$$\left. \begin{array}{c} x_U \\ x_L \end{array} \right\} = \bar{x} + \frac{b_1(y-\bar{y}) \pm ts\{(y-\bar{y})^2/S_{xx} + (b_1^2/n) - (t^2s^2/nS_{xx})\}^{1/2}}{b_1^2 - (t^2s^2/S_{xx})}$$

where t is the value from the $t(v,1-\alpha/2)$ distribution where v is the degrees of freedom of the sample variance (s^2). Liber et al. (1992) have suggested using this approach as a way to estimate the no effect concentration (NEC) by substituting the mean response prior to treatment as the y value. As noted by Finney (1978) the size of interval is determined by the sample size, the slope, and the sum of squares due to x. As in the above comparison of designs, the best approach is to spread the doses as far apart as possible. Thus again, design D1 with only two doses would produce the smallest interval.

Many researchers believe that two doses are not enough as the assumption about the linearity of the model may not be correct (Brown 1978, Giddings 1986). Another problem with using only two doses is that there is no protection against a nonlinear response. The two dose design (D1) is best when one knows the response is linear but may perform poorly when the response is nonlinear, as one is limited to fitting linear responses for this design (as there are only two doses used). The regression (simple linear) approach leads to a more powerful test if the true model is linear, and there are more than two doses. If the true model is not linear, then the power and ability to make inferences can be severely affected if a linear regression is fit (Smith and Mercante 1989). If one uses a more complicated regression model, the power to detect effects using an overall test of the regression model approaches that of the ANOVA model as terms are added to the regression. Data that are not linear arise frequently as it is not uncommon that low doses cause stimulation of processes such as growth and reproduction while greater doses cause inhibition.

This discussion of optimal designs, although useful, has some limitations. First, it works best if the model is correct. When the model is nonlinear, and the design is based on a linear model, the results will not be optimal and can lead to erroneous results (especially when using two dose levels). Second, the criteria used to compare the designs arise from models requiring normality for validity. When the normality of the data is not a valid assumption, the results may be misleading. For example, when the data are binomial, and interest is in low doses, and evenly spaced doses are used, it is advantageous to have more replications in the middle than at the ends (Krewski and Kovar 1982, Krewski et al. 1984, 1986). Third, other criteria may lead to different design approaches. The above approach assumes an upper and lower dose. The upper dose may not be known or may be based on laboratory data. In pesticide studies, the focus of the study may be on invalidating or validating a no effect dose from a laboratory study. For example, if the expected exposure level in the field is well below the LC_{50} from a toxicity study, this exposure level may be considered safe. In this case, a sensible approach (especially with a small number of samples) is to bracket the exposure dose and include a control. This approach would result in four doses and three replications per dose. The method, however, is dependent on the choice of the expected exposure level that is estimated using simulation (Graney et al. 1989).

Unfortunately, in the design of multispecies tests, there are additional problems, as emphasis is often on all of the above criteria and more, as the interests of researchers include multispecies hypotheses as well as single species hypotheses. Thus, the researcher is in considerable dilemma. To arrive at feasible designs for multispecies tests, prioritization and compromise are needed. A first step in the design is to list the reasons for the experiment. When there is a single objective (which is not very often), then an optimal design can be obtained. When there are multiple objectives, it is not possible to obtain the best design to address all the objectives, and compromise is needed. It is best to assign a priority to each objective and to try to find a design that will address the objectives with highest priority and will possibly address the low priority objectives.

Second, it is best to use as much information as possible on the chemical to aid in prioritizing objectives, setting levels, and evaluating designs. For example, if we know that a chemical primarily affects fish but does not affect plankton, then it is not sensible to design a multispecies test that focuses on plankton. The highest priority should be given to fish and secondary effects related to fish. It is much easier to design around primary effects than secondary effects (Crossland and LaPoint 1992). Also, information from other studies on different toxicants or chemicals could be used to aid design. This information is usually incorporated in a subjective fashion, although a more statistical approach is possible. Given priorities, objectives, endpoints, and estimates of variance and effects, it is then possible to compare some designs. There will not be a best design but some may be eliminated, and the limitations of others may become apparent. Also, the limitations due to sample size will become known. The limitations may lead to more realistic objectives and endpoints.

IMPLEMENTATION

Although the planning of the experiment is important, so too is the implementation of the study. A well-designed experiment is not safe from poor laboratory methods. It is important to control the dose levels to insure proper exposure. Attention must be paid to the measurement of the responses to insure that the measurement error is small. For example, if the sampled organisms are analyzed later or at a different site, it is important that the collection and transportation process does not increase greatly the variability in measurement.

It is important as well in multispecies tests to maintain the proper statistical perspective on collection of data. For example, in mesocosm studies it is important to note that the statistical unit is the mesocosm. While it may be useful to collect multiple measurements on each mesocosm at a given time, these measurements are made on the same unit and hence count as one statistical observation. For example, if interest is in how introduction of a toxicant alters the diets of fishes, the natural approach might be to look at the stomach content of individual fish. One approach to taking measurements is to empty each stomach, identify everything inside, and then create a data matrix based on individual fish and their diet. An alternate approach is to recognize that the unit is the group of fish in the mesocosm. Hence, we are not interested in individual fish but rather the group. Individual fish give information about variation between fish in the tank as well as a treatment effect. For statistical assessment, our desire is to estimate the treatment effect and control the variation between the fish. This may be accomplished by detailing the contents of a sufficient number of stomachs using the above scheme but can also be accomplished by collecting the stomachs, compositing the material inside, and then sampling from the composite. This approach will result in the same information if the number of fish is reasonably large but may reduce the cost.

ANALYSIS OF MULTISPECIES STUDIES

Just as there are a number of approaches for designing multispecies tests, so too are there a multitude of methods for analysis of the results. Data resulting from multispecies tests include information from few (fish) to many (zooplankton) species, often measured over time. The analyses are frequently difficult because of the large number of species and small sample sizes. In this section, several methods for the analysis of multispecies data are described. Some discussion is given to the limitations and benefits of different approaches. We do not attempt to cover all methods that are available for use; rather the focus is on methods not used in standard toxicity methodology (Rand and Petrocelli 1985). Three analytic approaches are discussed. First, some comments are given about the univariate approach. This approach is the most common and best documented. The second approach occurs when the test units are measured over time. Then, the experiment becomes a repeated measures study as multiple measurements are made on the same unit. The third approach is the multivariate approach.

Before discussing approaches, it is important to note that prior to analyzing data, the endpoint(s) must be chosen. The choices of endpoints define different classes of models that one may apply to the data. For example, if growth is the response, one may use ANOVA, regression, or nonlinear regression. When mortality is the response, categorical analysis and logistic regression may be appropriate. Endpoints also affect the view of the data analysis. Multivariate analysis may be applied when there is more than one endpoint of interest and interest is in a simultaneous analysis. Multivariate analysis may also be used to display the data and look for composites that are useful for separating effects. An approach that is sometimes used is to form the composite prior to analysis. Thus, the endpoint of the study becomes a measure that converts multivariate data to univariate data. For example, it is not uncommon to use the number of species or species diversity as endpoints to monitor in the study (Pratt et al. 1988, Pontasch et al. 1989).

THE UNIVARIATE APPROACH

In the univariate approach, there is a single endpoint, or, if there are several endpoints, each is considered separately. Methods for analyzing data differ depending on how one considers the time effect. The simplest approach is to consider each time separately or average across times. Then the problem is simply the analysis of either regression data or ANOVA data, depending on the hypotheses of interest (Liber et al. 1992). For example, if there is replication, then tests for an overall effect, tests to indicate no observable effects, and tests comparing different concentrations are possible. If interest is in a regression model, this can also be fitted. The regression model can be used to estimate quantities such as LC_{50}s or predict effects associated with different concentrations. As these methods are well described in the toxicity literature, there is little need to discuss them in detail here (see, for example, Finney 1978, Rand and Petrocelli 1978, Masters et al. 1991, Conquest and Taub 1989, or Weber et al. 1989). The most difficult question it seems in these analyses is the choice of endpoint. Environmentally appealing endpoints such as no observable effects levels are difficult to justify. As pointed out by Skalski (1981) and Stephan and Rogers (1985), NOELs often reflect model sensitivity and not a true "no effect". For there to be a true "no effect" in a regression model requires a threshold model (Cairns 1992). As regression models tend to be linear or curvilinear, the estimated no effect level is really a statistically insignificant dose. The estimates of these quantities may be greatly affected by the dose levels and sample sizes. This is true regardless of whether the estimate is based on regression (Liber et al. 1992) or ANOVA models. Unfortunately, sample sizes required to distinguish a threshold model from a linear regression may be prohibitive. A second difficult question is related to biological significance. With large sample sizes, differences may be detected that are not important biologically. It is useful in the design phase to pick a magnitude of effect that is biologically significant, then design the experiment to be able to detect that difference.

ANALYSIS OF REPEATED MEASURES

Many multispecies tests involve time as a factor. Dosing may be applied at the start of an experiment and the test unit measured at different times. Thus, the experiment becomes a two factor experiment (dose and time). There are two concerns that arise with these studies. First, time is not randomized. This may lead to problems with a univariate analysis of results. Second, each unit is measured over time. Measurements on the same unit are typically correlated (which violates the usual independence assumption). This type of experiment is referred to as a repeated measures experiment, and there are many approaches to its analysis (Crowder and Hand 1990).

One approach to analyzing this type of data is to focus on the time response curve for each concentration. This approach is useful if interest is in estimating recovery times or searching for optimum values (time when the chemical has the greatest effect). A complication in the analysis that must be considered is that the measurements made on the same multispecies test unit are correlated over time. The correlation generally makes the regression variance smaller than it would be if different units were used. Hence, for example, estimates of recovery times are perceived to be more precise than they really are.

A more complicated, but more realistic, approach is to analyze the data as a repeated measures experiment (Crowder and Hand 1990 is a good but difficult text; a simple text is Hand and Taylor 1987). The design discussed above involves two factors, dose and time. Dose is randomized to the units while time is not. In statistical experimental design books, the design is viewed as a split-plot design without blocking, while, in the psychology literature, it would be referred to as a mixed within and between subjects design (time is the within subjects factor and dose the between subjects factor). There are many ways to analyze this type of data.

A simple approach is to use the method of curves (Meredith and Stehman 1991). In this approach, a curve is fitted to each unit. For example, if there are 12 test systems, 12 curves are fitted. In fitting the curves, it is best to have equally spaced observations in time and use orthogonal polynomials. Then, the curves are viewed as replicates and can be analyzed using ANOVA methods to see if the curves change as a function of dose. An alternative approach is to view the measurements over time as a group of variables. Then, multivariate randomization methods can be applied to test for dose effects and test for time effects (Zerbe and Walker 1977, Manly 1991). By selecting contrasts on the time data (for example, comparing the controls within each time) randomization contrast tests can be performed (Foutz et al. 1985).

The second approach is to view the study as a two factor analysis of variance and use an ANOVA approach (Milliken and Johnson 1984). Then, one can investigate effects associated with the chemical, time, and the interaction between time and the chemical (Genter et al. 1987). It is possible to make multiple comparisons or carry out polynomial analysis of the data. One difficulty is that the follow-up analysis is complicated because the time factor is a repeated factor. This creates two problems. First is the problem that the unit creates one source of error, and the time measurement creates a second source of error. The existence of two error terms complicates the analysis of interaction means (see Milliken and Johnson, 1986 for details). The second problem is that the time measurements on the same units are correlated. This complicates testing for effects. Several approaches are available for making inferences. An approach leading to liberal tests is to ignore the problem and use the mean square error to test for time and the time–chemical interaction. A conservative approach is to adjust the degrees of freedom when obtaining the p-value from the F distribution. There are three common adjustments. The most conservative is to divide the degrees of freedom for time and the interaction by the degrees of freedom for time. The other approaches multiply by a factor between 1 and 1/(degrees of freedom for time). The two common methods for estimating this factor are called the Greenhouse-Gieser and the Hyunh-Feldt methods (see Crowder and Hand 1990). Generally, the Hyunh-Feldt method is considered the better method. Computer packages often print p-values associated with the unadjusted and the Greenhouse-Gieser and Hyunh-Feldt approaches.

If the above approach seems complicated or confusing, then consult a statistician. It can lead to complicated analyses. Another approach that also may seem complicated is to view the data as a multivariate set of data. Thus, the time measurements on each unit represent a vector of measurements. Then, the vectors of measurements are analyzed. As in the univariate analysis, there is focus on the chemical factor and the time factors. The same inferences discussed above can be carried out using multivariate methods (Capizzi and Burton 1978). One difficulty in using the multivariate approach is that the method is conservative. If the observations in time are uncorrelated or have a special correlation structure, then the analysis of variance tests are more sensitive. However, it is unlikely that the assumptions are met. The multivariate methods can be quite difficult to apply and interpret. Many extensions and variations on these methods are discussed in the statistics literature. Examples include adjustment for covariates, adjustment for time zero, repeated measurement analysis with categorical data, and multivariate repeated measures (Srivastava and Carter 1983, Crowder and Hand 1990).

MULTISPECIES ANALYSIS

Multispecies tests can also be analyzed using multivariate techniques by considering the species as variables. The difficulty in applying the techniques is that the sample sizes tend to be small relative to the number of species that are considered as the variables. Most multivariate methods are used as descriptive or graphical methods and include cluster analysis, principal components analysis, and correspondence analysis (Jongman et al. 1987). However, multivariate inferential methods may also be used. Crow and Taub (1979) use multivariate analysis of variance (MANOVA) to assess the effects of a toxicant. Another approach for a multivariate analysis is to use a randomization procedure. Smith et al. (1990) and Pratt and Smith (1991) used a similarity approach to analyze a multispecies study. The method was applied at a single time point, thus the method is analogous to a one way MANOVA. The data consist of species vectors, one vector of information for each unit. The approach is based on summarizing the information in the species data through a similarity matrix or distance matrix. Then the similarities are combined to form a statistic; example statistics would be the total of the similarities between species lists receiving different treatments.

The process involved in the similarity analysis method is illustrated in Figure 3. The first rectangle is the species array matrix. The rows of the matrix are the different units, divided according to level of chemical. Thus, there might be, for example, 4 levels of the chemical and 3 units per level giving a total of 12 test systems. The first step in the analysis is to summarize the data array as a distance or similarity matrix. Then, the matrix is reduced to a suitable statistic. If there is an effect associated with the chemical, the units receiving the same concentration of chemical should be more similar than two species lists that receive different concentrations of chemical. Thus, the distance matrix should have a pattern of high similarities (among units receiving the same concentration) and low similarities (between units receiving different concentrations). Thus a suitable statistic might be the mean between similarity (which should be low if there is an effect), the mean within similarity (which should be high), or the ratio (analogous to analysis of variance tests). If a distance measure is used, the same summary statistics could be used, but the interpretation is opposite (high between distance means an effect).

The second step is the permutation step. The species vectors are now permuted. Thus, for example, the first vector may become the third in the array, the third the tenth, etc. A new data matrix is formed that is the permutation of the rows of the original matrix. Again, this matrix is summarized using a distance or similarity measure. The second step is repeated a great number of

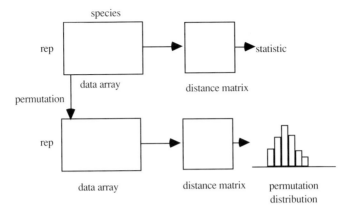

Figure 3   The randomization method for testing for a chemical effect using a species distance matrix approach.

times, say 1000, to obtain 1000 values of the statistic. If the hypothesis that there is no treatment effect is true, then the value of the statistic in the original data should be close to the permuted statistics. If, however, there is a chemical effect, then the original statistic should be considerably different from the permuted statistics. The permuted statistics can be summarized as a histogram or distribution and the location of the original statistic used to indicate significance. A p-value can be calculated as the number of permuted statistics that are greater than or equal to the observed statistic relative to the total number of permutations plus one (for the original value).

There are a variety of follow-up methods useful for further analysis and interpretation of results. Foutz et al. (1985) describe how the approach can be used to test contrasts or make multiple comparisons. Smith et al. (1990) describe some approaches for interpreting results. They give a method for evaluation of species importance and describe an approach for selection of a set of important species. Although it is possible to apply the randomization approach in a regression context, applications thus far have focused on testing hypotheses about group effects. Manly (1991) gives some examples of regression analysis using a randomization approach in other contexts. Pontasch et al. (1989) compared some of the approaches to assess macroinvertebrate response to a complex effluent.

The above tests provide a suite of methods for analyzing multispecies test data. The list is by no means complete, and there are other methods that may lead to informative analysis. Choice of method depends on the hypotheses of interest, the data that are collected, and the complexity of the design. The above methods primarily focus on direct effects (decreased abundance, loss of species, etc.). More difficult is the analysis of indirect effects. Indirect effects are sometimes of greater interest in multispecies tests, especially when the focus in ecological effects is not suggested by a single species test. These may include effects associated with changes in the food web, predator–prey relationships, and competition changes. These effects are more difficult to analyze because not only are appropriate tests required but also an understanding of the ecological processes and the underlying scales are often needed. A good interaction between ecology and statistics is essential to the understanding of these effects. Another difficulty in the analysis is the incorporation of other information. In the above discussion, the data are viewed as information on dose and the response to the dose. Measurements may also be taken on physical-chemical parameters, possibly at different temporal scales. Relating this information to changes in taxonomic infomation is not easy, especially with small sample sizes.

It is important to remember that all of the above statistical methods (including the randomization method) are based on assumptions. Thus, it is important to include checking the assumptions as part of the analysis of the data. Regression and ANOVA methods are based on assumptions of normality of the observations, homogeneity of variance, and independence of observations. Although the assumptions need not be strictly valid, it is worth the effort to make the checks and try to adjust for any problems.

SUMMARY AND COMMENTS

The design and analysis of useful multispecies experiments are challenging tasks. This chapter has described some of the approaches to design and analysis. There has been no attempt to make a thorough detailing of all the designs that may be used nor all the analyses that have been run on multispecies experiments. The reason they have been avoided is that there are a multitude of objectives, scales, and designs for studying the effects of chemicals on surrogate ecosystems. In designing studies, the differences in objectives and scales are important and need to be considered in the design.

Although multispecies tests are complicated by the number of objectives and the complexity of the system, it is important to consider a statistical approach to design. By detailing and prioritizing objectives, designs can be chosen that will increase the chance of meeting those objectives. Comparison of designs using different statistical criteria may not indicate an optimal design for all objectives but may lead to elimination of some designs and the narrowing of possibilities.

The analysis of multispecies experiments also presents some challenges. It is important again to keep objectives in mind and to perform analyses that address the objectives. Further, it is necessary that the analyses are appropriate for the design and the data. Designs that involve repeated measurement require special treatment to incorporate the correlations between measurements over time. It is important that the statistical assumptions be checked and appropriate adjustments made to see that they are met as closely as possible.

REFERENCES

Brown, C. C. 1978. The statistical analysis of dose-effect relationships. pp. 115–148 in G. C. Butler (ed.), *Principles of Ecotoxicology Scope 12.* John Wiley and Sons, New York.

Cairns, J., Jr. 1992. The threshold problem in ecotoxicology. *Ecotoxicology* 1:3–16.

Capizzi, T. and D. T. Burton. 1978. A statistical technique for analyzing time-response curves. pp. 137-149 in K. L. Dickson, J. Cairns, Jr., and R. J. Livingston (eds.), *Water Pollution Assessment: Quantitative and Statistical Analysis,* ASTM STP 652. American Society for Testing and Materials, Philadelphia.

Cochran, W. G. and G. M. Cox. 1957. *Experimental Designs.* John Wiley and Sons, New York.

Conquest, L. L. and F. B. Taub. 1989. Repeatability and reproducibility of the standard aquatic microcosm: Statistical properties. pp. 159–177 in *Aquatic Toxicology and Hazard Evaluation,* 12th volume, ASTM STP 1027, American Society for Testing and Materials, Philadelphia.

Crossland, N. O. and T. W. La Point. 1992. The design of mesocosm experiments. *Environ. Toxicol. Chem.* 11:1–4

Crow, M. E. and F. B. Taub. 1979. Designing a microcosm bioassay to detect ecosystem level effects. *Int. J. Environ. Stud.* 13:141–147.

Crowder, M. J. and D. J. Hand. 1990. *Analysis of Repeated Measures.* Chapman and Hall, London.

Dudzik, M., J. Harte, A. Jassby, E. Lapan, D. Levy, and J. Rees. 1979. Some considerations in the design of aquatic microcosms for plankton studies. *Int. J. Environ Stud.* 13:125–130.

Finney, D. J. 1978. *Statistical Method in Biological Assay.* MacMillan Publishing Co., New York.

Foutz, R. V., D. R. Jensen and G. W. Andersen. 1985. Multiple comparisons in the randomization analysis of designed experiments. *Biometrics* 41:29–38.

Giesey, J. P., Jr. and P. M. Allred. 1985. Replicability of aquatic multispecies test systems. pp. 18–247 in J. Cairns, Jr., (ed), *Multispecies Toxicity Testing,* Pergamon Press, New York.

Genter, R. B., D. S. Cherry, E. P. Smith, and J. Cairns, Jr. 1987. Algal-periphyton population and community changes from zinc stress in stream mesocosms. *Hydrobiologia* 153:261–275.

Giddings, J. M. 1986. Microcosm procedure for determining safe levels of chemical exposure in shallow-water communities. pp. 121–134 in J. Cairns, Jr. (ed.), *Community Toxicity Testing.* American Society for Testing and Materials, Philadelphia.

Giddings, J. M. and G. K. Eddelmon. 1979. Some ecological and experimental properties of complex aquatic microcosms. *Int. J. Environ. Stud.* 13:119–123.

Graney, R. L., J. P. Geisey, Jr., and D. DiToro. 1989. Mesocosm experimental design strategies: advantages and disadvantages in ecological risk assessment. pp. 74–88 in Voshell, J. R., Jr. (ed.), *Using Mesocosms to Assess the Aquatic Ecological Risk of Pesticides: Theory and Practice,* Miscellaneous Publication 75, Entomological Society of America, Hyattsville, MD.

Hand, D. J. and C. C. Taylor. 1987. *Multivariate Analysis of Variance and Repeated Measures.* Chapman and Hall, London.

Isensee, A. R. 1976. Variability of aquatic model ecosystems-derived data. *Int. J. Environ. Stud.* 10:35–41.

Jongman, R. H. G., C. J. F. ter Braak, and O. F. R. van Tongeren. 1987. *Data Analysis in Community and Landscape Ecology.* Centre for Agricultural Publishing and Documentation (Pudoc), Wangeningen, The Netherlands.

Kleinbaum, D. G., L. L. Kupper, and K. E. Muller. 1988. *Applied Regression Analysis and Other Multivariable Methods* (2nd Ed.). PWS-Kent Publishing Company, Boston.

Krewski, D. and J. Kovar. 1982. Low dose extrapolation under single parameter dose response models. *Commun. Stat. Simul. Computat.* 11:27–45.

Krewski, D., J. Kumar, and M. Bickis. 1984. Optimal experimental design for low dose extrapolation II. The case of nonzero background. pp. 167–191 in Y.P. Chaubey and T. D. Dwivedi (eds.), *Topics in Applied Statistics.* Concordia University, Montreal.

Krewski, D., M. Bickis, J. Kuvar, and D. L. Arnold. 1986. Optimal experimental designs for low dose extrapolation I. The case of zero background. *Util. Math.* 29:245–262.

Liber, K, N. K. Kaushik, K. Solomon, and J. H. Carey. 1992. Experimental designs for aquatic mesocosm studies: a comparison of the "ANOVA" and "regression" design for assessing the impact of tetrachlorophenol on zooplankton populations in limnocorrals. *Environ. Toxicol. Chem.* 11:61–77.

Manly, B. F. J. 1991. *Randomization and Monte Carlo Methods in Biology.* Chapman and Hall, London.

Masters, J. A., M. A. Lewis, D. H. Davidson, and R. D. Bruce. 1991. Validation of a four day *Ceriodaphnia* toxicity test and statistical considerations in data analysis. *Env. Toxicol. Chem.* 10:47–55.

Meredith M. P. and S. V. Stehman. 1991. Repeated measures experiments in forestry: focus on the analysis of response curves. *Can. J. For Res.* 21:957–965.

Milliken, G. A. and D. E. Johnson, 1984. *Analysis of Messy Data* Vol. 1. Van Nostrand Reinhold, New York.

Myers, R. H. 1976. *Response Surface Methodology.* Edwards Bros., Ann Arbor, MI.

Pontasch, K. W., E. P. Smith, and J. Cairns, Jr. 1989. Diversity indices, community comparisons and canonical discriminant analysis: interpreting the results of multispecies tests. *Water Res.* 23: 1229–1238.

Pratt, J. R., N. J. Bowers, B. R. Niederlehner, and J. Cairns, Jr. 1988. Effects of atrazine in freshwater microbial communities. *Arch. Environ. Contam. Toxicol.* 17:449–457.

Pratt, J. R. and E. P. Smith. 1991. Significance of change in community structure: a new method for testing differences. pp. 91–103 in *Biological Criteria: Research and Regulation,* EPA 440/5-91-005, Office of Water, U.S. Environmental Protection Agency, Washington, D.C.

Rand, G. M. and S. R. Petrocelli (eds.). 1985. *Fundamentals of Aquatic Toxicology: Methods and Applications.* Hemisphere Publishing Co., Washington D.C.

Skalski, J. R. 1981. Statistical inconsistencies in the use of no-observed effect levels in toxicity testing. pp. 377–387 in P. R. Branson and K. L. Dickson, (eds.), *Aquatic Toxicology and Hazard Assessment: Fourth Conference.* American Society of Testing and Materials, Philadelphia.

Smith, E. P. and D. Mercante. 1989. Statistical concerns in the design and analysis of multispecies microcosm and mesocosm experiments. *Tox. Assess.* 4:129–147.

Smith, E. P., K. W. Pontasch, and J. Cairns, Jr. 1990. Community similarity and the analysis of multispecies environmental data: a unified statistical approach. *Water Res.* 24:507–514.

Srivastava, M. S. and E. M. Carter. 1983. *An Introduction to Applied Multivariate Statistics.* North Holland Publishing Co., New York.

Stephan, C. E. and J. W. Rogers. 1985. Advantages of using regression analysis to calculated results of chronic toxicity tests. pp. 328–338 in R. C. Bahner and D. J. Hansen (eds.), *Aquatic Toxicology and Hazard Assessment: Eighth Symposium,* ASTM STP 891. American Society for Testing and Materials, Philadelphia.

Touart, L. W. 1988. *Aquatic Mesocosm Tests to Support Pesticide Registrations.* EPA-540/09-88-035. Hazard Evaluation Division, Technical Guidance Document. U.S. Environmental Protection Agency, Washington, D.C.

van Voris, P., D. A. Tollie, M. F. Arthur, and J. Chesson. 1985. Terrestrial microcosms: applications, validation and cost benefit analysis. pp. 117–142 in J. Cairns, Jr., (ed.), *Multispecies Toxicity Testing,* Pergamon Press, New York.

Weber, C.I. et al. 1989. *Short-term Methods for Estimating the Chronic Toxicity of Effluents and Receiving Waters to Freshwater Organisms.* Second Edition. EPA/600/4-89/001 U.S. Environmental Protection Agency, Washington, D.C.

Zerbe, G. O. and S. H. Walker. 1977. A randomization test for comparison of groups of growth curves with different polynomial design matrices. *Biometrics* 33:653–657.

CHAPTER 6

Development and Current Use of Single Species Aquatic Toxicity Tests

Donald I. Mount

HISTORICAL ROOTS

Single species toxicity testing of aquatic organisms is limited mostly to the 20th century. References to such testing are sparse in the earlier literature. The efforts in the first third of the century were largely confined to fish species and seemed to be motivated largely by scientific curiosity. Most testing involved natural characteristics of water such as dissolved oxygen, pH, temperature, and suspended solids. Of course, few of the chemicals of concern today were present in surface water of those days. Metals, probably both because of their natural presence as well as their occurrence in early discharges, were among the first "toxic" substances to be tested. Of course, methods were not developed for either testing or culturing, so, by today's standards, the work seems haphazard. Analytical methods for metals were based on wet chemistry that was hard to perform and subject to poor detection limits and many interferences. Therefore, measured concentrations were rarely reported. One must remember that, at that time, surface water was not impacted to any great extent from discharges, and, so, precise tolerance levels were really not of concern. Rather, the main intent was to observe how fishes were affected by such things as acid conditions.[1] There was no important need for careful exposure control, as we now try to achieve.

The middle third of the century marks a new phase and emphasis in aquatic toxicology. Cities were becoming large enough (and most lacked any waste treatment), that reaches of streams below discharges became anoxic from the decomposition of organic matter. This period was the heyday of stream biological survey work, documenting the degradation and recovery of aquatic

0-87371-599-3/95/$0.00+$.50
© 1995 by CRC Press, Inc.

communities from the influx of nutrients and depression of dissolved oxygen. Quite logically, workers recognized the need to know the low oxygen tolerance of organisms, especially fishes, and an increase in oxygen studies is seen in the literature.[2]

World War II resulted in a large growth of the steel industry and increased discharges of steel mill wastes. In both England and the U.S., one can see increased references to metal toxicity.[3] Concurrently, particularly Doudoroff in the U.S. was investigating what we now think of as effluent toxicity tests.[4] The establishment of the Taft Engineering Center in Cincinnati, Ohio, led to a world recognized center for aquatic toxicity testing during that period. Both Tarzwell and Doudoroff promoted aquatic toxicity testing while there.[5] Almost simultaneously, the development of DDT and then related chlorinated hydro-carbon pesticides, fueled both the interest and need for single chemical toxicity tests and the beginning of modern aquatic toxicity testing. Interestingly, testing of effluent toxicity did not grow, remaining on the back burner until about 1980, largely because of the emphasis on waste treatment in the Clean Water Act. When the available technology was in place, and effluents still had toxicity, interest in toxicity testing returned. Perhaps because of the larger industrial complex in the U.S., publications on aquatic toxicology by U.S. workers made up an increasing share of the literature of the period.

The practical need for aquatic toxicity testing for use in water pollution control programs was by far the dominant motivation for the development of the science. Even today, that dominance remains. In addition, data for registra-tion of new chemicals under a variety of laws was another important driving force for such testing, starting with pesticide testing requirements. This histori-cal summary would be incomplete without mention of the recent astronomical increase in aquatic testing for control of effluent toxicity. More testing is now done in a few weeks than was done *in toto* in the first third of the 20th century.

GENERIC TRENDS IN TEST METHODS

As mentioned above, the early testing used static methods necessitating tests of short duration. These methods remained in use until the middle of the century. Tarzwell was the strongest voice promoting flow-through testing and surely had a major influence in its development. Not only was he energetic in promoting such testing, but he was in charge of the aquatic group of the Taft Engineering Center that so dominated the field for a decade or so, at least in the U.S. The much more rapid progress in freshwater methods as opposed to marine methods, can clearly be attributed to Tarzwell's relentless efforts.

This group was also responsible for the emergence of the fathead minnow as a "standard" species in aquatic testing. At a State of Ohio hatchery near the Taft Center, fatheads were raised in ponds. These provided a ready source of test fish for testing the many chemicals studied by Tarzwell's staff. Not only because fathead minnows were adaptable to laboratory testing and widely available, but also because of the toxicity data base already developed, workers in other places began using them and their utilization increased.*

Working in Tarzwell's group in the 1960s and with his strong support, I developed methods for and completed the first full life cycle chronic tests using these fish.[6] With the impetus of events leading to Earth Day, flow-through testing and chronic tests were quickly adopted by most other groups as well, resulting in a much increased rate of data generation.

Two events followed later that again changed the entire field. McKim, working at the Duluth, MN, U.S. EPA National Water Quality Laboratory, proposed the approximate 30-day embryo larval tests as a surrogate for the much longer full life cycle tests.[7] His idea was rapidly accepted, and the use of full chronic tests diminished quickly. Not long after, the need for effluent tests emerged as the 1980s arrived. For large scale use, flow-through tests on effluents were not practical (and maybe not needed), and 30 days was clearly too long. These needs resulted in the emergence of the still current "mini chronic" era and the use of static renewal testing.[8] We thus came full circle back to essentially static testing.*

This full circle of events is not accidental. It is the result, again, of practical need. During the period when flow-though testing in full chronic tests was dominant, workers in the field were focused on the ultimate in data quality and maximum effect. The use of the data was largely for water quality standards. Registration of new chemicals was limited largely to pesticides and not much aquatic data was required there for a long time. For well treated effluents, non-persistent toxicity is less probable and therefore static or static renewal testing is usually adequate. Because effluent toxicity varies more than the difference between test methods, testing more samples with cheaper tests provides the most information for a fixed amount of resources.

The need for more and faster testing was driven by three events. One was the explosion of new synthetic chemicals. The second was increased requirements for aquatic toxicity data for pesticide, then for premarket registration under the Toxic Substances Control Act, the Food and Drug Act, and others. The third was the need for effluent toxicity tests. Anyone looking at the bigger picture could plainly see that the development of water quality standards could not keep pace with the new chemicals coming out or even catch up with the backlog. Furthermore, it was an after-the-fact way to control. Clearly the limited resources needed to be spent more effectively.

The sheer amount of needed testing meant shorter, cheaper tests. Some data on most chemicals would buy more environmental protection than much data on a few. At least the worst actors should be recognized. To the aquatic toxicologists credit, the trend in methods followed the most urgent need. We

---

*A humorous note — the Taft Center was dominated by sanitary engineers, and aquatic biologists were not their favorites. Tarzwell's group became affectionately called "Those fatheads at the Taft Center!"

thus come to our present situation and can examine the current use and abuse of single species toxicity tests with a framework in mind as to why they are so important now.

CHARACTERISTICS OF SINGLE SPECIES TEST DATA

The acute toxicity of chemicals to any organism, including aquatic organisms, has become rather standard information. Toxicity for each species–chemical combination is a reasonably reproducible value among laboratories and over time. The measurement of 48- or 96-h LC50 values, at least for environmental concerns, is basic information that is used for a host of purposes; as, for example, ranking chemicals into toxic groups, judging general degree of hazard for use and transport, as a first cut on wastewater discharge concerns, and making development and marketing decisions on new chemicals.

Because exposure conditions and durations have become standardized, the results of the test may be only weakly relevant to the conditions in the environment. This relevance is similar to a comparison of the solubility of a chemical at 25°C in distilled water to its solubility in natural water under natural conditions. In both cases one can expect discrepancies.

Much of the early criticism of single species data was directed against the concept of the acute test. In reality the test was acceptable, but the results were being used for purposes for which the data were not intended. Twenty or more years ago, I remember many criticisms regarding test animals: abnormal sensitivity caused by stress, improper diet, and intentional selection of overly sensitive test species. Any of these can be a valid criticism, but, largely, the problem has been a failure when using the data for natural conditions, to consider bioavailability, constancy of exposure, exposure duration, and loss of toxicant by fate processes such as volatilization or microbial degradation.

Exposure concentrations in most tests are constant, or at least that is the objective, whereas constant exposure is rarely the case in the natural situations. Water used in the tests is usually free of suspended solids and organic matter. These and many others are usually present in natural water and all or some may have extreme effects on the expression of toxicity. Perhaps because of emphasis in early studies on metal toxicity and toxicity effects of water hardness, hardness is very often selected to match natural conditions. However, suspended solids and organic matter are usually not managed in toxicity tests even though they have a greater effect on toxicity than does hardness, for most chemicals.

Only a few species are widely tested among freshwater organisms, and, in any one study, the use of more than two or three species is uncommon. Except for a few chemicals, there are data for very few species. Among the common species tested, daphnids and trout do seem to be, broadly speaking, more sensitive than other species. Fortuitously, information on more sensitive species has more often been obtained. This is not to say they are always sensitive or that they are the most sensitive, but, in general, the tested species are among the more sensitive ones. Perhaps species that tended to be less sensitive were of less interest and were dropped from use.

Mortality is by far the most common endpoint used. For environmental protection uses, something less than a lethal concentration is desired. This desire led to application factors and safety factors applied to LC50 data to reduce it to a nonlethal or no effect concentration. The same objective led to chronic tests to avoid the need to estimate safe concentrations. As mentioned above, the cost, monotony, and large amount of work required to do full chronic tests soon forced a reexamination of ways to predict or extrapolate chronic values from acute data (e.g., embryo-larval test and "mini chronic").

Test endpoints also shifted from death to growth and reproduction, endpoints still widely used. For a half century or more there has been an interest in the use of biochemical, behavioral, or pathological endpoints. These have never been accepted in any broad or consistent way, perhaps because they have not been accepted by the regulatory people. Industry people are not interested in their use either since the biological significance of these changes is not easily identified.

Currently, in effluent toxicity testing, the hypothesis test for statistically significant changes from an exposure has created many regulatory problems. Statistical significance alone is not a relevant basis for judging effluent toxicity and its real world meaning. Effort to find and use a better evaluation approach is increasing. Obviously, since variability increases around the threshold being measured, variability is not a logical basis on which to make statistical decisions because mean young production is of most real world importance and not individual variability. In two tests with the same mean young production, statistical difference may occur in one case and not the other because variability is different, yet the population effect is the same.

Aside from these problems of statistics, a surprisingly small amount of effort has been devoted to identifying biologically relevant endpoints, particularly for chronic data. One could hardly argue that 50% death is not biologically important, but a small reduction in growth or reproduction is quite another matter. The literature is replete with studies to compare the results of single species tests to either multispecies tests including several levels of biological organizations or to field studies, but rarely have such studies paid any attention to endpoints used or compared. For the single species test endpoints, the conventional ones have been used without comment.

The essence of the situation is that exposure duration, relevance of exposure concentrations used, water quality characteristics, endpoints, choice of species, and methods of data evaluation all are important considerations when using the single species data in the natural environment. Yet, because of the routine methods now in existence, single species toxicity test data are used as though they are always applicable to any situation. Therefore, it is the application of the data, not its inherent validity that is questionable.

IMPACT OF REGULATORY STRATEGY FOR EFFLUENT DISCHARGE AND CHEMICAL REGISTRATION

Because use of single species aquatic toxicity data to regulate discharges to surface water is so important and large scale, some specific comments are in order. These considerations are as relevant to single chemicals as they are to effluents.

Worst-case applicability is commonly followed for regulation. That is, the highest concentration for the longest duration under the worst natural conditions is usually assumed. Furthermore, additivity, if not synergism, is often assumed for mixtures. If effluent toxicity is at issue, an assumption of maximum toxicity at minimum dilution is nearly always chosen. Choosing the margin of safety is not a subject that is appropriately discussed here partly because it is a social decision. However, the margin of safety used has an important bearing on how well toxicity test data relate to actual conditions and the resulting impact.

There are perhaps two assumptions usually made by regulators that result in the most discrepancy between toxicity test data predictions and observed results. These are the assumptions that toxicants or toxicity do not degrade and that a constant exposure will be maintained for extended periods. Both of these conditions are uncommon in most natural situations, and these assumptions lead to large margins of safety. Bear in mind, the issue here is not whether the margin is too large, but rather how such choices will cause an apparent discrepancy between observed and predicted.

THE FUTURE

One does not need a very sophisticated crystal ball to see that as waste generation increases, the margin of safety will have to be reduced. This is true for all media—air, water, and land. Single species test data have done a remarkable job of protecting because, in my opinion, we have had such a large margin of safety. When and if those margins are reduced, we will then experience the true inadequacies of single species data. Then we will need far more sophisticated tests—tests like those discussed in other chapters of this book. Test conditions that much more closely duplicate natural conditions will be necessary. Degradation, lack of additivity (which is very common in effluents), fluctuating exposures, probability of critical life stage exposure, interspecies effects, and biological availability will have to receive careful evaluation.

As these events happen, single species toxicity tests will take their proper place which is a standard reference test, much like water solubility, vapor pressure, and octanol/water partitioning. Toxicity so measured will be part of the base data set that will be used to make early development and manufacturing decisions. However, when environmental regulation is implemented, much more relevant test conditions, species, and exposure patterns will have to be employed to assess accurately the tolerable exposure.

Single species tests have been the workhorse for a long period, and about the only workhorse. They should remain the workhorse for very early decision-making but should become part of a "multi-team hitch" as products go into production and widespread use. Single species data can be much more accurate if the issues discussed above regarding how the data are used would also be incorporated (i.e., exposure duration, degradation, etc.). However, if the margin of safety is reduced, then the deficiencies will make such data inadequate for specifying acceptable concentrations in natural environments.

The practitioners of multispecies testing must take some lessons and guidance from our experience with single species testing. For use in environmental regulations, multispecies data must be:

- Definitive
- Objective—not subjective
- Broadly applicable
- Rapid to obtain
- Able to be performed routinely

Any testing that meets these needs will err in a measurable number of situations, will not have the certainty that is most desirable, and will not satisfy the true scientist. The success of the methods used must be measured by the environmental improvement obtained and maintained and not by scientific satisfaction, however.

REFERENCES

1. Wells, M.M., Reactions and Resistance of Fishes in Their Natural Environment to Acidity, Alkalinity, and Neutrality, *Biol. Bull.* 29:221–257 (1915).
2. Alabaster, J.S., D.W.M. Herbert, and J. Hemans, The Survival of Rainbow Trout (*Salmo gairdnerii* Richardson) and Perch (*Perca fluviatilis* L.) at Various Concentrations of Dissolved Oxygen and Carbon Dioxide, *Ann. Appl. Biol.* 45(1):177–188 (1957).
3. Doudoroff, P. and M. Katz, Critical Review of Literature on the Toxicity of Industrial Wastes and Their Components to Fish: The Metal as Salts, *Sewage Ind. Wastes*, 25(7)802–839 (1953).
4. Hart, B., P. Doudoroff, and J. Greenbank, *The Evaluation of the Toxicity of Industrial Wastes, Chemicals and Other Substances of Fresh-water Fishes*, Philadelphia, PA: The Atlantic Refining Company (1945).
5. Tarzwell, C.M., Bioassay to Determine Allowable Waste Concentrations in the Aquatic Environment, *Proc. R. Soc. London. B.* 177(1048):279–285 (1971).

6. Mount, D.I. and C.E. Stephan, A Method for Establishing Acceptable Toxicant Limits for Fish—Malathion and the Butoxyethanol Ester of 2,4-D, *Trans. Am. Fish. Soc.*, 96(2)185–193 (1967).

7. McKim, J.M., Evaluation of Tests with Early Life States of Fish for Predicting Long-Term Toxicity, *J. Fish Res. Board Can.*, 34(8)1148–1154 (1977).

8. Mount, D.I. and T.J. Norberg, A Seven-day Life-cycle *Cladoceran* Toxicity Test, *Environ. Toxicol. Chem.*, 3(3):425–434 (1984).

CHAPTER 7

Are Single Species Toxicity Test Results Valid Indicators of Effects to Aquatic Communities?

Benjamin R. Parkhurst

INTRODUCTION

The context for this chapter is the use of single species toxicity tests to regulate the discharge of toxic substances to freshwater streams and lakes, which is regulated under the National Pollutant Discharge Elimination System (NPDES) program of the Clean Water Act (CWA). Why should anyone care about this issue? The reasons for caring are both scientific and regulatory. In the opinion of some individuals, if an effluent is toxic, then it should not be discharged to surface waters under any circumstances. However, virtually all chemicals are toxic to some aquatic species given a sufficiently high dose, concentration, and duration of exposure. Therefore, under such a stipulation, effluents would not be discharged to surface waters regardless of whether or not they had any adverse effects to surface waters. The real scientific issue is whether the effluent poses a significant risk to aquatic life in the receiving water. This concern is incorporated into most state water quality standards, which generally include narrative standards that prohibit the discharge of toxic substances in toxic amounts. These narrative standards are interpreted to prohibit the discharge of chemicals in amounts that would be toxic to aquatic organisms in receiving waters. At present, the U.S. Environmental Protection Agency (U.S. EPA) enforces these narrative standards by placing whole effluent toxicity (WET) limits on effluents, without actually assessing or measuring effects to the aquatic community. The *Technical Support Document for Water Quality-Based Toxics Control*[1] (TSD) states that "Permit limits that are devel-

0-87371-599-3/95/$0.00+$.50
© 1995 by CRC Press, Inc.

oped correctly from whole effluent toxicity tests should protect aquatic biota if the discharged effluent meets these limits." Conversely, it can be inferred that, if these limits are exceeded, impacts to aquatic biota are likely to occur.

WET LIMITS

Typically, compliance with WET limits is judged by periodic monitoring of WET, such as on a weekly or quarterly basis. Tests are run using one or two species, generally a fish and an invertebrate. In some cases, tests with a plant species are required.

Acute tests are of 48- to 96-h duration, and the toxicity endpoint is survival, while chronic tests are of 7-days duration, and the endpoints are survival and growth for fish and survival and reproductive success for invertebrates. Plant tests are of 96-h duration, and the endpoints are effects on primary production.

Exceedences of WET limits in NPDES permits generally are indicated by the measurement of any statistically significant toxicity, using prescribed test methods, at an instream waste concentration (IWC) that would occur given the maximum effluent discharge volume and at a low stream flow, such as the 7Q10 stream flow. The 7Q10 flow is the 7-day average low flow with a recurrence interval of once in 10 years. Toxicity in chronic tests generally is judged at the designated IWC, while toxicity in acute tests may be judged in 100% effluent, or at some IWC.[1]

The question this chapter attempts to address is: Are single species WET test results valid indicators of significant toxic risks to aquatic communities? If WET test results are valid indicators of such effects, then significant WET measured at the IWC should be indicative of adverse toxic effects to the aquatic community. If significant WET is not measured, then adverse toxic effects from WET should not be occurring.

CONCERNS

In evaluating the utility of single species WET tests, the primary scientific and regulatory concerns are that use of single species WET tests may lead to an unacceptable incidence of false positive or false negative measures of toxicity. In other words, WET tests may overestimate or underestimate the actual effects that occur to aquatic communities. If WET tests overestimate actual effects, then, as a result of such test results, dischargers may be forced to incur costs to reduce the toxicity of their effluents, the result of which would be of negligible benefit to the aquatic community in the receiving water. In contrast, if WET tests underestimate actual effects, then, as a result of such test results, dischargers may continue to discharge effluent to receiving waters that are causing toxic effects to aquatic biota, when they are assumed not to.

I define a false positive as WET test results that show significant toxicity, when toxic effects to the resident biological community in the receiving stream are not occurring. A false negative is WET test results that show no significant toxicity, when toxic effects to the resident biological community in the receiving stream are occurring.

False positives and negatives may result because of (1) weak relationships between single species WET test results and responses of aquatic communities to effluent toxicity, and (2) the inherent variability of WET tests.

What constitutes adequate validation of the utility of single species WET test results for indicating effects to aquatic communities? I propose that single species toxicity tests are valid for predicting the toxicity of effluents to aquatic communities, if the frequency of false positive and false negative measures of toxicity is acceptable. However, the definition of an acceptable frequency of such outcomes is a policy decision, not a scientific one.

Given these proposed definitions, I will review available data on studies that have attempted to validate WET tests for predicting toxic effects to aquatic communities. Then, I will discuss some of the reasons why WET test results may sometimes overestimate or underestimate actual effects to aquatic communities.

VALIDATION STUDIES

The U.S. EPA[1] has focused on four data sets as evidence that WET test results are valid indicators of effects to resident aquatic communities: the U.S. EPA's Complex Effluent Toxicity Testing Program (CETTP) studies, the South Elkhorn Creek, Kentucky study,[2] a North Carolina study by Eagleson et al.,[3] and the Trinity River, Texas study.[4]

The eight CETTP studies include Scippo Creek, Ohio,[5] Ottawa River, Ohio,[6] Skeleton Creek, Oklahoma,[7] Five Mile Creek, Alabama,[8] Ohio River, West Virginia,[9] Kanawha River, West Virginia,[10] Naugatuck River, Connecticut,[11] and Back River, Baltimore Harbor, Maryland.[12]

The CETTP, South Elkhorn Creek, and Trinity River studies all assessed relationships between ambient (e.g., surface water) toxicity and ecological effects. Study sites ranged from small effluent dominated streams to large rivers (e.g., the Ohio River) with relatively small percentages of effluent. The Eagleson et al.[3] study assessed the relationship between WET and the benthic macroinvertebrate community.

The eight CETTP studies were designed "to investigate whether or not an evaluation of effluent toxicity, when adequately related to receiving water conditions, (i.e., temperature, pH, salinity), can give a valid assessment of receiving system impacts on waters that support aquatic biota."[1] In the 1985 edition of the TSD, the CETTP studies were the U.S. EPA's[13] primary scientific justification for the use of WET limits in NPDES permits. The U.S. EPA[13] concluded that "...at each of the eight sites investigated to date,...when exposure and potential chemical interactions are assessed, effluent toxicity can be correlated directly to impact in receiving waters."

However, Parkhurst et al.[14] pointed out that the CETTP studies did not investigate the statistical relationship between effluent toxicity and either instream (i.e., ambient) toxicity or biological community effects in the receiving stream. Therefore, this conclusion could not be directly derived from the CETTP studies.

In the 1991 edition of the TSD, the U.S. EPA[1] no longer concludes that the CETTP results and those of Birge et al.[2] and Dickson et al.[4] show correlation between effluent toxicity and effects to biological communities; rather, the U.S. EPA discusses the relationship between ambient toxicity and instream effects and the need to consider properly the actual dilution of the effluent in the receiving stream when predicting effluent toxicity. The U.S. EPA[1] now concludes that "The results (of the studies), when linked together, clearly show that if toxicity is present after considering dilution, impact will also be present."

The CETTP, South Elkhorn Creek, and Trinity River studies were relatively similar in that each study included measurements of ambient surface water toxicity using single species tests [*Ceriodaphnia* spp. and/or fathead minnows (*Pimephales promelas*)] and measurements of aquatic community structure (fish, benthic macroinvertebrates, algae, and/or zooplankton) in the receiving water. The toxicity tests measured effects on survival, growth, and/or reproduction, while the aquatic community studies generally were based on measurements of species richness.

A variety of quantitative and qualitative methods have been used to compare the toxicity test and aquatic community results. In general, the analyses showed that there was often a significant relationship between ambient toxicity and ecological effects. However, the relationship is variable and dependent on the type of analyses done, as I will illustrate below.

Dickson et al.[4] determined the significance of relationships between ambient toxicity test results and effects to the aquatic community in the CETTP, South Elkhorn Creek, and Trinity River studies. Among these studies, the proportion of false positives found using this approach ranged from 7 to 11%. They concluded that a strong relationship between ambient toxicity and effects to the aquatic community existed in the examined data sets. Dickson et al.[4] cautioned that their results were derived from studies where high levels of ambient toxicity were present and suggested that, where ambient toxicity was low or marginal, it will be difficult to elucidate a relationship because of the confounding effects of biological variability, habitat, or other sources of toxicity.

Using the CETTP and South Elkhorn Creek data, Parkhurst et al.[14] calculated (1) simple correlations comparing measures of ambient toxicity (i.e., *Ceriodaphnia* young per female and fathead minnow growth) to indicators of community effects (i.e., the number of taxa per taxonomic group: benthos, fish, periphyton, and zooplankton) and (2) canonical correlations for each study, which jointly consider all of the preferred measures of ambient toxicity and community structure. In these nine studies, among the 51 individual correlation analyses, significant relationships were found in only 11 (22%, Table 1). Five of eleven (45%) of the studies had at least one statistically significant simple correlation; also, 45% of the studies showed a statistically significant relationship between ambient toxicity and community effects using canonical correlation. Therefore, using this approach, in 45% of the studies, when significant ambient toxicity was measured, some kind of response in the aquatic community was also measured.

The South Elkhorn Creek study is an example where significant correlations were observed between ambient toxicity and community effects. All of the measured response variables followed similar trends at most stations (Figure 1). It is important to note that this close relationship between ambient toxicity and community effects was based on toxicity tests with fathead minnows only and that (1) both effluent and ambient toxicity were demonstrated to be highly lethal to the test organisms at actual IWCs; and (2) the study stream was effluent dominated, comprised of 60 to 100% municipal wastewater treatment plant effluent that contained acutely toxic concentrations of residual chlorine.

Marcus and McDonald[15] also evaluated the CETTP, South Elkhorn Creek, and Trinity River data to determine the capabilities of these ambient toxicity tests to predict correctly that no impacts to resident aquatic biota were occurring in receiving streams where none were found. Figure 2 illustrates some of the data they evaluated. Among the three studies, 13.8% of the sites did not have observed impacts to the aquatic community, yet 68% (0.094/0.138) of these were predicted by the ambient toxicity tests to be impacted. While Marcus and McDonald[15] suggested that these values might be used to suggest there was a relatively high percentage of false positive predictions of effects to resident biota from the results of ambient toxicity tests, they stated that this would be a misuse of these data. Similarly, these data cannot be used to support the general applicability of these single species test methods.

The State of North Carolina study compared WET test results with effects to benthic macroinvertebrates in a number of streams.[3] No statistical correlation between effluent toxicity and effects to benthic macroinvertebrates was done in this study; only a qualitative comparison was made, i.e., when the effluent was found to be toxic, were impacts observed to benthic macroinvertebrates, or were they not observed? In the State of North Carolina study, the presence of WET toxicity corresponded qualitatively with effects to benthic macroinvertebrates at 88% of the sites. However, for those sites at which no impacts to macroinvertebrates were found, the WET tests incorrectly predicted effects at 23% (0.07/0.30) of the sites (see Figure 1 to 3 in the TSD[1]).

The importance of these rates of false positive predictions of instream impacts is that the sites selected for these validation studies were not selected

Table 1 Summary of Correlation Analyses Between Ambient Toxicity and Numbers of Community Taxa (+ = Significant at P ≤0.05, 0 = Not Significant at P ≤0.05. Blank Cells Indicate That the Parameters Were Not Measured)

Study	Ceriodaphnia, No. Young/Female				Fathead Minnow, Mean Weight			
	Benthos	Fish	Periphyton	Zooplankton	Benthos	Fish	Periphyton	Zooplankton
Scippo Cr.	0	0						0
Ottawa R.								
1982	0	0	0					
1983	0	0						
Five Mi. Cr.								
February	0	0	0		0	0		
October	0	0			0	0	0	
Skeleton Cr.	0	0		0	0	0		0
Ohio R.	+			0	0	0		0
Naugatuck R.	+	+	+	+	0	0	0	0
Kanawha R.	+	0						0
Back R.	+	0		0	+	+		0
S. Elkhorn Cr.[a]					+	+		

[a] Because mean weight was not reported, the correlation analysis used fathead minnow, percent survival.

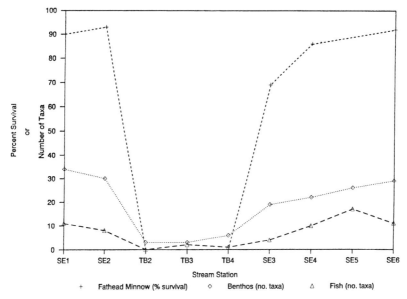

Figure 1   Comparison of ambient toxicity test results and numbers of biological community taxa at each sampling station for the South Elkhorn Creek, Kentucky
study.[2] The effluent was discharged upstream of stream station TB2. SE1 and
SE2 are upstream of the effluent discharge. Station TB2 and SE6 are downstream of the effluent discharge.

randomly and, therefore, do not represent surface waters in general.[15] Most of
the sites for these studies were selected because impacts were thought to be
present. Thus, it would be incorrect both scientifically and statistically to
extrapolate these results to predict effects at other sites, unless they are very
similar, hydrologically and ecologically, to the study sites discussed above.

The incorrect predictions of effects in these studies may have been due to
the following:

1. the presence of site-specific factors, such as habitat, ambient water quality,
   flows, or others that confounded the toxic effects of the water to resident
   aquatic biota;[14] and/or
2. the measurement of significant toxicity in a sample when the sample was not
   truly toxic; this could be caused by the inherent variability of the tests.

CONFOUNDING FACTORS

The potential effects on community structure of factors other than ambient
toxicity generally were not considered in the validation studies discussed
above. The authors of these reports implicitly assumed, therefore, that effluent
or ambient toxicity, as defined by the toxicity tests used, was the only factor
affecting the numbers of community taxa in the study streams. While I also

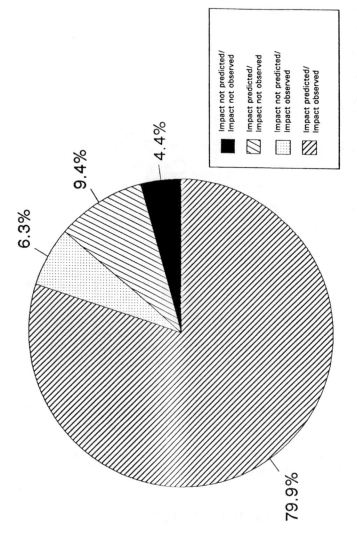

Figure 2 Comparison of ambient toxicity and instream impact—CETTP, Trinity River, and South Elkhorn Creek studies. (From U.S. EPA Technical Support Document for Document for Water Quality-Based Toxics Control, 1991.)

agree that effluent toxicity was an important factor affecting aquatic life toxicity at some of these sites, it was not the only factor. The lack of a consistent correlation between ambient toxicity and responses of the aquatic community may have resulted, in part, from other factors, both natural and anthropogenic, that influence the condition of biological communities. In some cases, these other factors may reduce the numbers of biological taxa in the same manner and, at the same stations, as does toxicity (e.g., discharges of toxic effluents may be spatially associated with other impacts on the stream system). Such situations would increase the probability of observing a significant correlation among ambient toxicity and community effects, when no direct cause-and-effect relationship actually exists.

Factors present in the South Elkhorn Creek, North Carolina, and CETTP studies identified in a review by Parkhurst et al.[14] that could have confounded interpretation of the data on biological community structure include poor habitat, low instream dissolved oxygen (DO), nutrient enrichment, organic enrichment, nonpoint sources of toxicity, and acute toxicity episodes (Table 2). All of these factors can have the same type of qualitative effects as effluent toxicity on the number of community taxa. The measures of ambient toxicity and community status collected in the studies could not and did not distinguish between reductions in taxa caused by effluent or ambient toxicity and those caused by other factors. Further, in most cases, the potential importance of these other factors was not recognized or evaluated. For example, numbers of fish and benthic macroinvertebrate taxa may be decreased by low DO levels. Low DO may be caused by nutrient and organic enrichment, possibly originating from the same effluent that was evaluated for toxicity. For example, Martin et al.[16] found DO depletion in the Ottawa River associated with effluent discharges, which was the site of the CETTP study. If these DO depletions also occurred during the CETTP studies (DO depletions were not studied in the Ottawa River CETTP study), the reductions in benthic macroinvertebrate and fish taxa reported in the study may have been partially or totally caused by low DO.

Acute toxicity episodes were detected in several of the CETTP studies. Such episodes, even if infrequent, could have a greater effect than chronically toxic effluents on the distribution and abundance of species, if the episodes were relatively severe. Lewis et al.[17] reported that an acutely toxic episode occurred in the dilution water collected above the sewage treatment plant on the Ottawa River in Lima, Ohio, resulting in a significant reduction in *Ceriodaphnia* survival on day 4 of a 7-day test. Such events were reported to be common and their impact significant in controlling the biological quality of the river.

An additional potential confounding factor occurring in all of the ambient toxicity tests was the unknown influence of the natural food (e.g., detritus, bacteria, algae) present in the stream water samples on *Ceriodaphnia* reproduction and

Table 2　Toxicological, Biological, and Habitat Factors That Were Identified in Each Study, Which May Have Confounded the Relationships Among Effluent Toxicity, Ambient Toxicity, and Community Effects

Parameter	Scippo Cr.	Ottawa R.	Five Mile Cr.	Skeleton Cr.	Ohio R.	Naugatuck R.	Kanawha R.	Back R.	S. Elkhorn Cr.	N. Carolina
Toxicological										
Ambient Food	X	X	X	X	X	X	X	X		
Atypical Response		X				X				
Fungal Growth	X						X			
High Salinity				X				X		
Low Dissolved Oxygen			X	X		X	X	X		
Methodological Variability						X				
Nonpoint Sources of Toxicity		X	X							
Replicate Variability					X					
Temporal Variability		X	X							
Toxic Episodes		X				X	X			
Biological										
Methodological Variability	X									
Temporal Variability		X	X							
Habitat										
Flocculent Material	X			X						
Hydrological Variability	X	X	X	X	X	X	X	X	X	
Low Dissolved Oxygen		X							X	
Metal Precipitates	X									
Nutrient Enrichment	X	X		X						
Organic Enrichment/BOD		X		X			X	X	X	X
Overhead Cover				X					X	
Physical Variability	X	X	X		X	X	X	X	X	
Salinity			X	X				X		
Sedimentation	X							X		
Low Temperature										

fathead minnow growth. Decreases or increases in available food can cause the same types of effects as toxicity, i.e., reductions or increases in numbers of young and growth. Variations in food availability seem to be particularly important for the *Ceriodaphnia* tests. In many of the CETTP studies, the reference stations (i.e., those with the greatest measured levels of *Ceriodaphnia* reproduction and fathead minnow growth) were downstream of effluents, where natural supplies of food may have been increased due to nutrient or organic enrichment. One example of this is the Skeleton Creek CETTP study, where the station farthest downstream (Station 9) was the reference station because it had the greatest production of young, higher even than Station 2 upstream of all effluents (Figure 3). The effects of different levels of natural, ambient sources of food on the toxicity test results were not evaluated in these studies. Therefore, decreases in reproduction or growth resulting from ambient toxicity could not be separated from the effects of variations in food supplies.

The potential importance of diet in these tests is highlighted by the study of Lewis et al.[17] on the Ottawa River at Lima, Ohio, in 1986. Lewis et al.[17] reported that the combination of yeast, CEROPHYL, trout chow (YCT) diet, a diet formula that is nutritionally improved over that used in the CETTP studies, and reconstituted water used in the *Ceriodaphnia* tests appeared to be nutritionally inadequate. Interestingly, this inadequacy was not apparent when *Ceriodaphnia* were raised in the river water. Lewis et al.[17] hypothesized that the adequacy of the YCT diet was dependent upon the nutrient level and bacteria content of the dilution water.

Another source of error in predicting effects to aquatic communities is the large degree of variability present in measurements of aquatic community structure. Much of this variability is due to the inherent variability of aquatic communities, as well as sampling error and measurement error. This variability may confound the measurement of adverse effects present in communities and, therefore, contribute to false negative measurements of toxicity (i.e., predictions of no adverse effects, when they are, in fact, present). In such communities, the degree of effects has to be large before they can be determined to be statistically significant.

VARIABILITY OF SINGLE SPECIES TESTS

As previously discussed, an additional cause of false positive and false negative predictions of toxicity to aquatic communities from single species test results would be over- or underestimates of the toxicity of the sample caused by the inherent variability of the tests.

Like all environmental measurements, measurements of toxicity test endpoints exhibit variability. Uncertainty in the test endpoint decreases the reliability and confidence in the measured outcome. Sources of uncertainty in toxicity tests include inter- and intralaboratory variation in individual test organism sensitivity, culturing techniques and diet, implementation of test

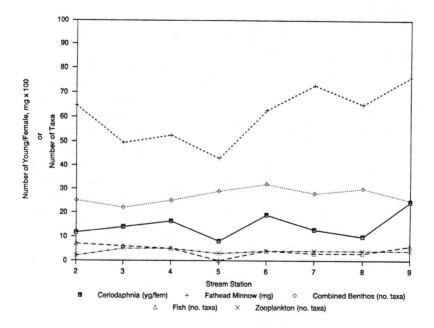

Figure 3 Comparison of ambient toxicity test results and numbers of biological community taxa at each sampling station for the Skeleton Creek, Oklahoma, CETTP study.[7] Effluents were discharged to Skeleton Creek between stream stations 2 and 3 and between stream stations 4 and 5.

methods, data recording, analytical procedures, and others. It is important to evaluate this uncertainty in interpreting toxicity test results.

The example data that I will use to illustrate the uncertainty in WET test results are the results of an interlaboratory study by DeGraeve et al.,[18] which analyzed 7-day mortality of fathead minnows. Toxicity tests were conducted on two different chemicals and three different effluents at each of 10 laboratories. For each test material, the data consisted of four replicate tests per concentration conducted at each laboratory, for a potential sample size of 40 tests per concentration. Each test consisted of six different concentrations. Because some tests were not completed, the actual sample size ranged from 34 to 38 tests per concentration (see Table 3–1 in DeGraeve et al.[18]).

Figure 4 provides an example of the probability of false positive (and false negative) measures of toxicity for the 7-day fathead minnow test compared to an assumed toxicity limit of >50% mortality. Let us assume that the predicted probability of survival (solid line), estimated by applying logistic regression to the raw data, is the best estimate of the true toxicity of the effluent. The column of dots on the far left side of the figure illustrates the percent survival at each of the laboratories for the controls [pentachlorophenol (PCP) concentrations = 0.0 µg/L]. The next three columns of dots illustrate the range of survival at the

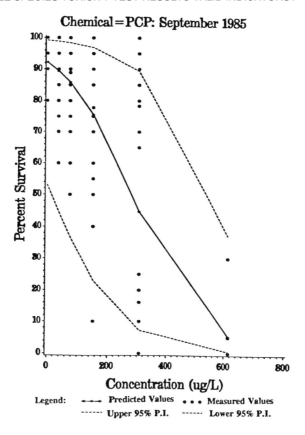

Figure 4   Inter- and intralaboratory percent survival using the 7-day fathead minnow data with pentachlorophenol (PCP), September 1985.

three lowest test concentrations of PCP, 38.5, 77.0, and 154.0 μg/L. If we assume that PCP was present in an effluent or ambient surface water sample at these concentrations, and no other toxic chemicals were present, we would expect a similar range in survival among the ten laboratories for the samples. The predicted probability of survival for effluent with PCP concentrations at these three levels is >50%; therefore, the effluent should have passed the toxicity limit. However, there were at least two false positive measures of toxicity (note: since some test results at different laboratories were the same, some dots represent results from more than one laboratory). For the next higher concentrations of PCP, 308 μg/L, the predicted probability of survival was <50%; therefore, the effluent should have failed the toxicity limit. There would have been at least eight false negative measures of toxicity. Only at 616 μg/L PCP were there no false positive or false negative measures of toxicity. Similar rates of potentially false positive and false negative measures of toxicity were found in other data sets, including tests with single chemicals and effluents and tests with fathead minnows, *D. magna* and *C. dubia*.[19] Therefore, the inherent variability in such tests can lead to a significant proportion of false positive and false negative indications of effluent toxicity, which would lead to incorrect predictions of effects to aquatic communities.

CONCLUSIONS

The question this chapter started out to address was: Are single species WET test results valid indicators of significant toxic risks to resident aquatic communities? However, as I discussed previously, of the four validation studies used by the U.S. EPA[1] to test this relationship, only one, the North Carolina study,[3] actually investigated this relationship, and this study only looked at qualitative relationships in effluent dominated streams in one state, North Carolina. The preponderance of the validation studies evaluated relationships between ambient surface water toxicity and effects to biological communities. Consequently, the validity of the utility of single species WET tests for predicting effects of effluents to aquatic communities has not been adequately tested.

In these validation studies, measurements of ambient toxicity based on single species tests provided good indications of higher level community effects, under some conditions. These conditions were situations in which ambient toxicity was relatively high. These conditions tended to occur in streams that were effluent dominated, i.e., a high proportion of the stream was comprised of effluent, and the effluent was relatively toxic. In situations where ambient toxicity was low or marginal, confounding factors such as biological variability, community resilience, habitat, other sources of toxicity, and the variability of single species tests tended to obscure or overwhelm any effects toxicity may have had on the resident community.

In these validation studies, single species test results were not shown to be useful for predicting specific kinds or degrees of effects; rather, they were shown to be useful for indicating whether effects to aquatic communities should or should not occur, given that significant toxicity was or was not measured in surface waters. For example, a certain percent survival, growth rate, or number of young produced in a single species toxicity test was not predictive of specific effects to an aquatic community, such as a percent decrease in numbers of macroinvertebrate taxa. However, if significant, relatively severe, chronic toxicity was measured in the stream, then it was likely that some type of measurable effect had occurred within the aquatic community.

As Parkhurst and Mount[20] have discussed, these validation studies demonstrated that, if proper consideration was made for the dilution of the effluent in the receiving water, WET should be a reliable predictor of ecological effects. However, in these validation studies, care was taken to compare the results of toxicity and field response at comparable concentrations and exposure times.

These studies were also done in such a way that decay of toxicity was included in the comparison, or time was short enough that decay did not occur.

The regulatory applications of WET limits in most instances, however, ignore decay of toxicity, and, in many instances, limits are imposed at such low dilution critical flow choices that treatment requirements more stringent than necessary to protect aquatic life, and designated uses of receiving waters are required. This situation is further aggravated because duration of exposure is essentially ignored in regulatory use, and a finding of toxicity above the prescribed limit is treated as though it may have existed for a long period. Toxicity limits are based on a minimum dilution, but, when allowable toxicity is exceeded, the reaction to it is never viewed in light of the dilution that existed at the time of occurrence.

The IWC for toxicity limits (and chemical-specific limits) in NPDES permits is typically not based on the actual IWC that occurs when the test is run but is generally based on an IWC calculated from the maximum effluent discharge volume and a minimum stream flow, such as the 7Q10. Limits are based on effects seen at low flow effluent concentrations to provide a margin of safety (i.e., if effects are not observed at low flow concentrations when effluent concentrations are relatively high, then they will not occur at higher stream flows when effluent concentrations are less). The significance of this is that the 7Q10 flow is a very low probability event; it occurs an average of 0.2% of the time. In addition, the maximum effluent toxicity is assumed to occur concurrently with the 7Q10 stream flow. However, the probability of these two events occurring concurrently is far less than 0.2% of the time.

Since the dilution at which the toxicity limit is set in a NPDES permit is, on the average, exceeded 99.8% of the time in the receiving water, the probability that WET will be correlated with effects to resident aquatic organisms is small. In streams that are not effluent dominated and that have highly variable flows, the difference between the actual IWC and that specified in the NPDES permit normally will be quite large, while, in effluent dominated streams, the difference between the IWC specified in the NPDES permit and the average IWC will not be so great. Therefore, only in streams that normally are comprised of a high proportion of toxic effluent, will the WET limits allowable in the NPDES permit be approached or occur a significant proportion of the time, and, consequently, effects to the biological community in the receiving water may be expected when the WET limit in the permit is exceeded.

Consequently, some degree of uncertainty is present in using single species toxicity test results to predict effects to aquatic communities. This uncertainty appears to be caused by factors such as variability in habitat, hydrology, water quality, effluent toxicity, biological communities, and single species test results. Therefore, it is vital to assess the relative contributions of these factors to such predictions and their potential roles in causing false positive and false negative predictions. The presence of uncertainty, however, does not preclude use of single species tests to monitor and regulate toxicity. Rather, it only demonstrates that uncertainty must be considered in the application and interpretation of single species toxicity tests. Consequently, if biological, toxicological, and ecological variability are considered properly, single species toxicity tests on ambient water samples have valid utility for indicating effects to aquatic communities. This relationship is strongest when levels of ambient toxicity are high and weakens as the level of toxicity decreases.

REFERENCES

1. U.S. Environmental Protection Agency, *Technical Support Document for Water Quality-Based Toxics Control,* Washington, DC: U.S. Environmental Protection Agency, Office of Water (1991), 313 pp.
2. Birge, W.J., J.A. Black, T.M. Short, and A.G. Westerman, A Comparative Ecological and Toxicological Investigation of a Secondary Waste Water Treatment Plant Effluent and Its Receiving Stream, *Environ. Toxicol. Chem.* 8:1373–1403 (1989).
3. Eagleson, K.W., D.L. Lenat, L. Ausley, and F. Winborne, Comparison of Measured Instream Biological Responses with Responses Predicted Using the *Ceriodaphnia* Chronic Toxicity Test, *Environ. Toxicol. Chem.* 9:1019–1028 (1990).
4. Dickson, K.L., W.T. Waller, J.H. Kennedy, and L.P. Ammann, Assessing the Relationship Between Ambient Toxicity and Instream Biological Response, *Environ. Toxicol. Chem.* 11:1307–1322 (1992).
5. Mount, D.I. and T.J. Norberg-King, Validity of Effluent and Ambient Toxicity Tests for Predicting Biological Impact, Scippo Creek, Circleville, Ohio, U.S. EPA 600/3–85/044 (1985), 86 pp.
6. Mount, D.I., N.A. Thomas, T.J. Norberg, M.T. Barbour, T.H. Roush, and W.F. Brandes, Effluent and Ambient Toxicity Testing and Instream Community Response on the Ottawa River, Lima, Ohio, U.S. EPA 600/3–84–080 (1984), 97 pp.
7. Norberg-King, T.J. and D.I. Mount, Validity of Effluent and Ambient Toxicity Tests for Predicting Biological Impact, Skeleton Creek, Enid, Oklahoma, U.S. EPA 66/8–86/002 (1986), 50 pp.
8. Mount, D.I., E.A. Steen, and T.J. Norberg-King, Validity of Effluent and Ambient Toxicity Tests for Predicting Biological Impact on Five Mile Creek, Birmingham, Alabama, U.S. EPA 600/8–85/015 (1985), 107 pp.
9. Mount, D.I., T.J. Norberg-King, and A.E. Steen, Validity of Effluent and Ambient Toxicity Tests for Predicting Biological Impact, Naugatuck River, Waterbury, Connecticut, U.S. EPA 600/8–86/001 (1986), 122 pp.
10. Mount, D.I. and T.J. Norberg-King, Validity of Effluent and Ambient Toxicity Tests for Predicting Biological Impact, Kanawha River, Charleston, West Virginia, U.S. EPA 600/3–86/006 (1986), 73 pp.
11. Mount, D.I., A.E. Steen, and T.J. Norberg-King, Validity of Ambient Toxicity Tests for Predicting Biological Impact, Ohio River, near Wheeling, West Virginia, U.S. EPA 600/3–85/071 (1986), 60 pp.

12. Mount, D.I., A.E. Steen, and T.J. Norberg-King, Validity of Effluent and Ambient Toxicity Tests for Predicting Biological Impact, Back River, Baltimore Harbor, Maryland, U.S. EPA 600/8–86/001 (1986), 97 pp.

13. U.S. Environmental Protection Agency, *Technical Support Document for Water Quality-Based Toxics Control,* Washington, DC: U.S. Environmental Protection Agency, Office of Water (1985), 133 pp.

14. Parkhurst, B.R., M.D. Marcus, and L.E. Noel, Review of the Results of EPA's Complex Effluent Toxicity Testing Program, Washington, DC: Utility Water Act Group (1990), 181 pp.

15. Marcus, M.D. and L.L. McDonald, Evaluating the Statistical Bases for Relating Receiving Water Impacts to Effluent and Ambient Toxicities, *Environ. Toxicol. Chem.* 11(10): 1389–1402 (1992).

16. Martin, G.L., T.J. Balduf, D.O. McIntyre, and J.P. Abrams, Water Quality Study of the Ottawa River, Allen and Putnam Counties, Ohio, Ohio Environmental Protection Agency Report OEPA 79/1 (1979), 65 pp.

17. Lewis, M.A., W.S. Eckhoff, and J.D. Cooney, Impact of an Episodic Event on the Toxicity Evaluation of a Treated Municipal Effluent, *Environ. Toxicol. Chem.* 8:825–830 (1989).

18. DeGraeve, G.M., J.D. Cooney, T.L. Pollock, N.G. Reichenbach, and J.H. Dean, Precision of the EPA 7-Day Fathead Minnow Larval Survival and Growth Test, Intra- and Interlaboratory Study, Electric Power Research Institute Report EA-6189 (1988), 123 pp.

19. Warren-Hicks, W. and B.R. Parkhurst., Performance Characteristics of Effluent Toxicity Tests: Variability and Its Implications for Regulatory Policy, *Environ. Toxicol. Chem.* 11:793–804 (1992).

20. Parkhurst, B.R. and D.I. Mount, The Water Quality-Based Approach to Toxics Control: Narrowing the Gap Between Science and Regulations, *Water, Environ., Technol.* 3:45–47 (1991).

CHAPTER 8

Naturally Derived Microbial Communities as Receptors in Toxicity Tests

B. R. Niederlehner and John Cairns, Jr.

INTRODUCTION

Models are the most basic of scientific tools. When an entity is too large to manipulate, too valuable to risk, or impossible to replicate, the relevant characteristics of the system may be reproduced in a more manageable form. In the field of ecotoxicology, many microcosms have been designed to reduce relevant characteristics of natural ecological systems to tabletop size.[1]

Models are not exact reproductions of the systems being studied as they incorporate only the parts needed. O'Neill and colleagues[2] cite the example of the irrelevance of the color of a pendulum in describing its motion. Some microcosm toxicity tests exploit designed differences in naturally covarying gradients of complexity, time, and size (Table 1). In a small, fast, yet functionally complex system, the effects of toxic substances on endpoints, such as biotic diversity and stability, can be evaluated quickly. Characteristics of ecosystems that are valued and worthy of protection are used to evaluate the ecological integrity of natural systems. Analogous measurements in easily replicated and controlled model systems are essential to establish cause and effect. However, the success with which the relevant properties of ecosystems have been included in microcosms is debated. Microcosm test observations must be extrapolated across spatiotemporal and phyletic scales in order to predict socially relevant ecological outcome in natural systems. These extrapolations must be carefully evaluated.

Table 1   Covarying Scale Gradients in the Design and Application of Toxicity Tests

SPATIOTEMPORAL SCALE:	LOW ⟷ HIGH	
Small experimental units (e.g., test container)	⟷	Large experimental units (e.g., rivers, lakes)
Small receptors (e.g., bacteria)	⟷	Large receptors (e.g., fish)
Receptors with limited vagility	⟷	Receptors with great vagility
Receptors with short generation times (e.g., bacteria)	⟷	Receptors with long generation times (e.g., fish)
Endpoint early in progression of impairment (e.g., biomarker)	⟷	Endpoint of obvious social and biological importance (e.g., fisheries harvest)
Short observation time in relation to receptor's life history	⟷	Long observation time

DESIGN COMPLEXITY:	LOW ⟷ HIGH	
Few components	⟷	Many components
Standardizable	⟷	Site-specific
Repeatable	⟷	Realistic background variation
Capable of high precision	⟷	Capable of high accuracy
Low cost per experimental unit	⟷	High cost per experimental unit

BIOLOGICAL SCALE:	LOW ⟷ HIGH	
Endpoints low in levels of organization (e.g., survival of individuals)	⟷	Endpoints high, integrating effects at lower levels (e.g., nutrient retention of system)
Receptors low in taxonomic hierarchy (e.g., procaryotes)	⟷	Receptors high (e.g., vertebrates)
Receptors low in food web (e.g., primary producers)	⟷	Receptors high, integrating effects at lower levels (e.g., piscivores)

CORRESPONDENCE BETWEEN ENDPOINT AND ENVIRONMENTAL VALUE: LOW ⟷ HIGH		
Bottom-up appraisal of risk	⟷	Top-down appraisal of risk
Cause known	⟷	Environmental outcome known
Many tests are combined to provide information to model environmental effects (many experimental units)	⟷	Few tests are used to predict environmental effects (few experimental units)
Model used to translate test endpoints to prediction of environmental effects (with error)	⟷	Information directly applied to site-specific prediction (with error)

The artificial substrate microcosm (AS-M) is one of many microcosms designed to incorporate increased complexity into evaluations of toxicity. Although similar in spatiotemporal scale and design to conventional toxicity tests,[3] the community-level endpoints monitored in the AS-M are more similar to those used to evaluate natural systems.[4] This discussion summarizes our accumulated experiences with the AS-M: endpoint selection, repeatability, representativeness, and predictive validity of the AS-M to date.

DESCRIPTION OF TEST

BASIC DESIGN

A standard procedure for the AS-M is described by Pratt and Bowers.[5] A general description with some recent modifications follows. In the AS-M, microbial communities are obtained by colonizing small blocks ($6 \times 5 \times 3.75$ cm) of reticulated polyurethane foam (PF) for 7 to 21 days in a reference system or in a control location for a site-specific evaluation. The interstices of the artificial substrate are colonized by indigenous microbes; primarily by bacteria, fungi, algae, protozoans, and rotifers. Communities are then collected and transported to the laboratory for use in designed experiments.

As in conventional toxicity tests, methods of toxicant delivery can vary, and both ambient and effluent dilution tests have been performed. Tests can be static, static with replacement, or flow-through at various rates. The experimenter can control toxicant exposure, extent of replication, light intensity, photoperiod, current, and diluent water quality. In addition, invasion pressure can also be controlled — the test tank or diluent headbox is stocked with additional substrates, providing an "upstream" source of recolonists.

In a typical dilution toxicity test, three replicates of five concentrations and a control are tested. Experimental units are 4 to 8 L tubs. A field-colonized substrate is suspended at the head of each tub. One to six similar, but uncolonized, substrates are suspended at the opposite end of the tub (Figure 1). Test media flow into the tub, around the field-colonized substrate, around the barren substrate, and out of the tub. The test monitors colonization, i.e., the establishment of communities on initially barren substrates. Tests are monitored for 7 to 21 days. Comparisons of results after 7 days and 21 days (Table 2) suggest that, in most cases, response is similar.

Because the biological receptor in the AS-M is a community rather than a single organism, many endpoints can be monitored beyond reproductive success of individuals. Some of these endpoints correspond closely to societal goals. For example, because biodiversity is an environmental value worthy of protection, tests directly monitoring the effects of stress on biodiversity are particularly valuable. By directly monitoring a characteristic of interest, the increasing uncertainty and error involved in extrapolating from one easily measured endpoint to another with relevance to management goals[6] is reduced.

Responses that may be general symptoms of stress across different types of ecosystems and across different types of stress have been suggested.[7-9] We have monitored related endpoints in the AS-M (Table 3). Both signal-to-noise ratios (i.e., the sensitivity, Table 4) and consistency of response over different toxicant types (i.e., the generality, Table 5) of various endpoints have been compared. The most consistently useful endpoints (but not necessarily the most sensitive in any particular test) have been those that focus on characteristics of the community as a whole, not on individual functional or structural components.

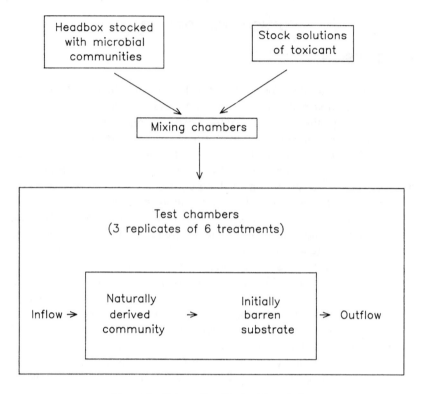

Figure 1   Schematic of test system design.

Table 2   Comparison of Effect Levels (IC20s for Losses in Taxonomic Richness) after 7- vs. 21-day Observation Periods

Toxicant	7 day	21 day
Ammonia	44	13
Cadmium	2.2	0.9
Chlorine	9.2	2.7
Copper	16.5	14.5
pH	7.7	6.0
Selenium	54	73
Zinc	15	20

Taxonomic richness and composition have been both consistent and sensitive indicators of stress. Changes in net ecosystem production (i.e., the biomass accumulation measures and net daily metabolism) were also consistent but not always among the most sensitive endpoints. Other endpoints were quite sensitive for individual toxicants but did not respond consistently across toxicants (e.g., algal biomass measures for zinc vs. ammonia).

Table 3   General Stress Syndrome: Common Features of Response to Stress Across Ecosystem Type and Related Endpoints Monitored in AS-M

Characteristics in General Stress Syndrome	Endpoints in AS-M
Loss of sensitive species, decreased species diversity	Loss of taxonomic richness, diversity indices
Increased instability in component populations, greater amplitude of fluctuations	Comparisons of treatment group variance over time
Changes in the size spectrum	Mean size of dominant taxa
Retrogression to opportunistic species	Similarity dendrograms over colonization
Lifespans decrease, turnover of organisms increase	Turnover rate, generation time of dominant taxa
Increased disease and parasite prevalence, decreased condition	Adenylate ratios, pheophyton/chlorophyll ratios
Trophic dynamics shift, food chains shorten, functional redundancy declines	Autotrophic index, net daily metabolism, loss of predators, trophic group analyses
Changes in primary production	Production to biomass ratios
Decreased efficiency of nutrient recycling, increased loss of nutrients	Nutrient pool size, nutrient uptake/regeneration rates
Efficiency of energy use decreases, respiration increases	Total production, respiration to biomass ratios
Increased circulation of contaminants	Toxicant fate analyses

Loss of taxonomic richness has proven particularly useful in extrapolating effects observed in the AS-M to different spatiotemporal scales. Taxonomic loss is a prominent part of field evaluations of biotic integrity[4] as well as the conceptual basis for models used to derive water quality criteria.[10,11] In addition, loss of taxonomic richness has been used to model toxicant effect in integrated environmental assessments.[12] Finally, loss of taxonomic richness is directly relevant to the societal goal of maintaining biodiversity. Taxonomic richness has strong ties to other environmental assessment methods and consistent discriminatory power, so it will be the focus of further discussion.

Data analyses for the AS-M are completely analogous to those in conventional toxicity tests. No-observed-effect-concentrations (NOECs) for each endpoint can be identified through Dunnett's tests (or nonparametric Steele's many-one tests). Regression analysis is used to derive concentration-response models. Inhibitory concentrations (ICx where x is the percent impairment in a quantitative response compared to the control) are inversely predicted from these models. Although taxonomic richness is a discrete variable, it can be treated as a continuous variable for the purpose of analysis because values are not far apart and cover a wide range.[13] Because the tolerance of various organisms to toxic stress is generally thought to be normally distributed (with a few very sensitive species, a few very tolerant species, and the great majority responding somewhere in between), log-logit (or log-probit) concentration-response regressions are used to model effects.[14] The degree of impairment that constitutes a biological response meriting a management response depends on at least four factors: the magnitude of natural background variation (noise), the

Table 4   **Signal-to-Noise Ratios for Endpoints Monitored in the AS-M over Eight Toxicants.[64] Noise is the Mean Minimum Significant Difference Detectable. Signal is the Maximum Difference Observed over Concentrations Spanning Two Orders of Magnitude. Both Signal and Noise are Expressed as % of Control Response**

Response	Noise	Signal	Ratio
BIOMASS:			
Ash Free Dry Weight	61.1	160.0	2.62
Total Protein	118.1	471.7	3.99
Total Organic Carbon	44.1	85.5	1.94
ATP	58.5	92.5	1.58
NUTRIENT POOL SIZE:			
Total Phosphate	39.3	72.7	1.85
Potassium	32.5	28.1	0.86
Magnesium	14.1	16.8	1.19
Calcium	10.9	10.4	0.95
Nitrate	140.8	89.1	0.63
Nitrite	22.8	16.2	0.71
Sulfate	30.9	22.0	0.71
Fluoride	37.3	32.4	0.87
NONTAXONOMIC STRUCTURE:			
In Vivo Fluorescence	89.1	277.0	3.11
Chlorophyll a	231.7	312.6	1.35
Hexosamine	139.9	149.4	1.07
Bacterial Counts	12.7	17.8	1.40
Pheophytin/Chlorophyll ratio	19.9	23.4	1.18
METABOLISM:			
pH	2.3	2.7	1.17
Dissolved Oxygen	6.4	8.8	1.38
Alkaline Phosphatase	259.2	664.3	2.56
ETS	164.7	412.3	2.50
Spectral Complexity	24.4	38.1	1.56
P/R	12.6	4.3	0.34
TAXONOMIC STRUCTURE:			
Species Richness	24.4	51.1	2.09

magnitude of change with stress (signal), the shape of the concentration-response curve, and potential for early warning of cumulative effects. These factors differ with the endpoint. Losses in taxonomic richness with a 20% impairment are a reasonable compromise between natural variability and minimal impact. Minns and colleagues[12] also use a 20% impairment in taxonomic richness as an index of biologically significant effects.

The person-hour requirements and approximate costs of an AS-M monitoring loss in taxonomic richness, net daily metabolism, mean size of dominant organisms, net system production, assimilative efficiency, and respiration-to-biomass ratio after a 21-day colonization are described in Table 6. The cost of $3,321 can be compared to the cost of a similar community-level prediction of environmental outcome based on arrays of conventional toxicity test data.[11] Based on a cost of $1,500 for short-term, chronic tests with single species, a

Table 5 Consistency of Response to Stress over Eight Toxicants. Table Entries are Ranks of the Treatment Intensity at which a Significant Difference from Controls was First Observed in the AS-M[64]

Response	NH$_3$[a]	Atr[b]	OCl[c]	STP[a]	Ind[e]	pH[f]	Phe[g]	Zn[h]
BIOMASS:								
Ash Free Dry Weight	>5	–[i]	>5	5	3	–	>5	1
Total Protein	>5	5	4	5	2	>5	>5	1
NUTRIENT POOL SIZE:								
Total Phosphate	–	–	–	PDG[j]	–	>5	–	1
Potassium	5	>5	5	–	–	>5	>5	–
Magnesium	–	3	>5	–	–	>5	>5	–
Calcium	–	3	>5	–	–	>5	>5	–
NONTAXONOMIC STRUCTURE:								
In Vivo Fluorescence	>5	–	4	>5	<1	>5	>5	2
Chlorophyll *a*	>5	5	3	4	–	–	–	–
Hexosamine	–	>5	–	5	–	>5	>5	–
Bacterial Counts	>5	–	–	4	>5	5	>5	>5
Pheophytin/Chlorophyll ratio	>5	–	–	4	–	–	–	–
METABOLISM:								
pH	–	–	>5	PDG	5	PDG	>5	5
Dissolved Oxygen	1	3	3	PDG	2	>5	4	1
Alkaline Phosphatase	–	–	2	4	–	–	–	4
TAXONOMIC STRUCTURE:								
Species Richness	2	5	2	5	3	4	>5	4
Composition	<1	2	<1	<1	2	<1	–	<1

[a] NH$_3$ = ammonia, [b] Atr = Atrazine, [c] OCl = chlorine, [d] STP = sewage treatment plant effluent, [e] Ind = industrial effluent, [f] pH = pH, [g] Phe=phenol, [h] Zn = zinc, [i] – = not monitored, [j] PDG=preexisting dilution gradient.

community-level prediction using three to seven of these tests[11] would cost $4,500 to $10,500.

ADVANTAGES OF THE AS-M

Microbial Communities as Receptors

If cost and efficiency were not factors, all components of any community affected by stress would be monitored. However, this is not generally practical, either in the field or in a complex microcosm. Instead, most evaluations rely on sentinel groups of organisms or on integrative endpoints with environmental or economic importance.[15] Although the environmental importance of microbial communities is enormous[16,17] and their influence on the sustainability of economic activities demonstrable,[18] microbial species are not directly valued by laymen. For this reason, damage to microbial communities alone may not provide a convincing demonstration of environmental harm for the general public or its political representatives. However, microbial communities, especially protozoans, may be a uniquely informative group to test.

First, the number of protozoan taxa tested simultaneously in the AS-M is large. In a typical test, 20 to 80 taxa of protozoans will occur on a control

Table 6   Costs and Time Required for a 21-day AS-M. Estimate Does not Include Set Up Costs (Microscope, Taxonomic Keys, Training) or Costs for Analytical Confirmation of Toxicant Concentrations

Person hours	
Planning	8
Preliminary Preparations	4
Test setup/maintenance	40
Measuring endpoints	24
Data analyses	16
	92
Salaries @ $16/h	$1,472
Fringe benefits @ 30% of salaries	$ 442
Materials and supplies	$ 200
Computer/instrumentation pro rata	$ 100
Direct costs	$2,214
Indirect costs @ 50% of direct costs	$1,107
Estimated cost	$3,321

substrate at any one time, and over 200 taxa will be represented over the entire experiment.[19] Instead of conducting 30 tests to determine effects on 30 taxa, one test provides that information. This lends a certain efficiency to determinations of range in response to toxicants.[20] Because the numbers of taxa exposed are large and replication is easy, a 20% change in species numbers is easily detected. While it may be difficult to assess the importance of the absence of one of five fish taxa, or two of ten macroinvertebrate taxa, a decline of 10 of 50 protozoan taxa is usually statistically significant.[19]

Second, common protozoans are functionally diverse. Picken[21] observed that co-occurring protozoans are not merely a collection of individuals relying on the same food source, but are an interacting complex of herbivores, carnivores, omnivores, and detritus feeders. As a consequence, effects of toxicants that are specific to a particular trophic style (e.g., photosystem inhibitors on primary producers or biomagnifying substances on predators) will directly affect some but not all protozoan taxa. Indirect effects resulting from elimination of a keystone species or trophic category can be detected. Microbial communities include top predators, and, on their own spatiotemporal scales, suctorians and lake trout are equally important in shaping their communities. Their presence will shape the outcome of toxic stress in their respective systems. The elimination of either without replacement by functionally similar species will drastically alter their community because a trophic guild is eliminated. Trophic cascades[22] have been detected in microbial communities through analysis of taxonomic and trophic guild composition.[23] It has also been shown that alterations in trophic state in the microbial community correspond to trophic changes in macroinvertebrate communities operating at a larger spatiotemporal scale.[24]

   Third, protozoans have unique phyletic diversity. Small and Green[25] suggest that there are 46 or more protistan phyla. Recent molecular phylogenies of eukaryotic organisms suggest that protists occur in half of the major clades of higher eucaryotes and in all but one clade for lower eucaryotes.[26] This phyletic diversity is useful in toxicity testing because a diverse group is most likely to characterize the range of sensitivity to toxic stress that can be expected in aquatic organisms. Closely related species have similar structural targets and physiological mechanisms; however, increasing taxonomic distance makes response to toxicants less predictable.[6,27,28] While algae, but not fish, are sensitive to photosystem inhibitors such as atrazine, and arthropods, but not fish, are sensitive to chitin inhibitors such as diflubenzuron, some protozoans directly respond to each. Similarly, responses to metals are ameliorated by metallothionein in game fish[29] and common protozoans.[30] However, there are limits to this generality. Pratt and colleagues[31] found that microbial communities were unaffected by chlorpyrifos concentrations up to 0.12 µg/L, but this concentration was sufficient to affect the acetylcholinesterase activity of organisms with nervous systems. In a progression of impact, physiological changes are expected at lower stress levels than community level responses; however, since microbes lack the same target for chlorpyrifos, a nervous system, any affects seen in the AS-M must be expressed through a different target.

   In general, do small organisms respond differently from large ones? Some evidence indicates that the tolerance distributions from the AS-M are not biased. Estimates of the concentration of chemical stress adversely affecting 20% of organisms based on log-logit, concentration-response models of tolerance from the AS-M and from arrays of conventional toxicity test data with fish, crustaceans, insects, algae, snails, etc. are compared in Table 7. For 8 of 11 chemicals, the ratios of the maximum to the minimum IC20 are <3 and, so, comparable to variation in repeated conventional toxicity tests.[32] However, for the organic chemicals phenol and pentachlorophenol, effect levels in the AS-M are substantially greater than those from collections of larger species tested individually. Why? Since these chemicals are readily biodegraded, available concentrations within the substrate may have been lower than measured water column concentrations. Alternately, while taxonomic richness often increases with mild organic enrichment in natural systems, organisms tolerant to organic enrichment have been deliberately excluded from toxicity test method development in the now abandoned search for "the most sensitive species". Standard test species may not provide a random sample of tolerances to organic enrichment.

   As a final point supporting the use of microbial communities in toxicity tests, it is interesting to note that, of all the taxonomic groups studied in the manipulation of whole lakes in the Experimental Lake Area studies, it was the small, quickly reproducing, easily dispersed microbial species that provided

Table 7   Comparisons of Permissible Chronic Levels
Based on Log-logit Concentration-response
Models of Tolerance from AS-M and Arrays of
Single Species Tests. Table Entries are IC20s
Followed by the Number of Genera on Which
Estimates are Based

Toxicant	AS-M	Single Species	Ratio Max/Min
Ammonia	13	18	1.35
(unionized µg/L)	31	9	
Atrazine	274	93	2.94
(µg/L)	37	6	
Cadmium	0.9	1.6	1.92
(µg/L)	34	10	
Chlorine	2.7	4.1	1.52
(µg/L)	41	3	
Copper	14.5	8.5	1.71
(µg/L)	30	12	
pH	6.0	6.3	2.00
(units)	41	9	
Pentachlorophenol	165[a]	8.3	19.88
(µg/L)	57	18	
Phenol	9500	112	84.82
(µg/L)	62	6	
Selenium	73[b]	49	1.49
(µg/L)	62	3	
Trinitrotoluene	251[a]	80	3.14
(µg/L)	63	6	
Zinc	20	56	2.80
(µg/L)	34	6	

[a] Pratt and colleagues[31]

[b] Pratt and Bowers[5]

the most consistent early indicators of stress.[9] If it is impractical to describe entire communities, protozoans may provide an unusually inclusive representation of target structures, mechanisms, interactions, and possible responses in the environment.

The Use of Naturally Derived Communities

When naturally derived communities and their associated substrate are used in toxicity tests, both the individual species and the organization of the community are indigenous. The successional process has "designed" a system. In contrast, a gnotobiotic mixture of species assembled in the laboratory is not coselected and may not constitute a system with feedback mechanisms and homeostasis.[2] Intact feedback mechanisms may affect background variability. Microbial communities on PF substrates exhibit characteristics of intact organization in their homeostatic behavior, such as the ability to maintain stable and characteristic numbers of protozoan species in laboratory systems over time periods sufficient for a number of complete turnovers of organisms.[33] For example, communities developed in experimental units receiving three levels

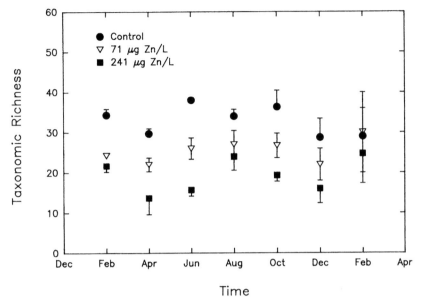

Figure 2 Variation of taxonomic richness over time and over treatment group.

of zinc for over a year had stable and distinguishable numbers of taxa[34] (Figure 2). Variations in taxonomic richness within each zinc treatment over time were small ($p = 0.1350$), while treatment differences were significant ($p = 0.0039$). The coefficient of variation for taxonomic richness in control communities over a year was 14.4%. This can be compared to coefficients of variation generally >50% and varying from 20 to 200% over 64 days for structural endpoints in gnotobiotic microcosms.[35]

Balance Among Observational Scales

The AS-M coordinates relative scales of receptor and system size, length of observation, hierarchical scale of endpoints, and intensity of stress.[36] Spatial scale is defined in terms of size or vagility of the organisms observed and temporal scale in generation times or turnover time for the assemblage.[33]

The size of the interstices of the PF substrate limits the size of inhabitants to those <500 μm. Larger, slower, more free-ranging organisms that would have infrequent but gross effects on individual communities are size excluded. As a consequence, the frequency of many relevant ecological processes is circumscribed and the timeframe of interest is defined. Table 8 describes the size, generation time, and taxonomic and functional complexity of groups within the microbial communities. The ratio of observation time to the longest generation time for the major constituents is 1:3. To obtain the same ratio, a macroinvertebrate community with generation times from weeks (in multivoltine dipteran species) to >1 year (for many trichopterans) would have to be

Table 8   Relative Scales of Biological Components in a Typical Microbial Community on a PF Artificial Substrate

Group	Generation/ doubling time (day)	Number per ml	Size (largest dimension in μm)	Number of taxa	Functional groups[a]
MAJOR COMPONENTS:					
Bacteria	<1	500,000	1	>4,000[b]	D
Algae	<1	200,000	5–300	20–80	A, 1
Protozoa	<1	20,000	5–300	20–80	A, D, 1, 2
Fungi	1–7[c]	(1.0[c])	500	ND[d]	D
Rotifers	2–7	1–4	75	1–12	D, 1, 2
MINOR OR TRANSIENT COMPONENTS:					
Oligochaetes	2–25	0–3	500	1	D, 1
Nematodes	3–21	0–4	500	ND	D, 1, 2
Cladocerans	3–21	0–2	300	1	D, 1
Gastrotrichs	3–22	0–3	200	ND	D, 1
Copepods[e]	17–365	0–3	75	1	D, 1
Ostracods	14–365	0–3	500	1	D, 1, 2
Insects[e]	14–365	0–1	500	1	D, 1, 2

[a] A = autotroph, D = detritivore, 1 = primary consumer, 2 = secondary consumer

[b] Based on Torsvik and colleagues[65]

[c] Based on hexosamine concentrations in μg/ml

[d] ND = no data

[e] Limited to early life stages

studied for 1 to 3 years. The ratio may facilitate comparisons between scales. For example, recovery times for microbial communities under ideal conditions, i.e., a good upstream source of colonists and no lasting effects on habitat chemistry or physical structure, are about 7 days.[37,38] This is one generation time of the longest-lived, major constituents. In macroinvertebrate communities, recovery typically takes 1 to 3 years.[39] The ratio of recovery to generation time is, again, 1.

The ratio of the size of the individual to experimental unit size in the AS-M is also similar to that in much larger test systems. Most organisms on the substrate are ~10^{-12} cm³; the substrate itself is ~10^2 cm³; the tank in which the substrate is suspended is ~10^4 cm³. The difference between the individual and the experimental unit size is 16 orders of magnitude. Compare this to a small fin-fish (~10^2 cm³) in a pond mesocosm (~10^8 cm³). The difference between the individual and the experimental unit size is six orders of magnitude.

Inclusion of Invasion Pressure

Because the AS-M includes invasion pressure, species susceptible to toxic stress and those that avoid it can be replaced by functionally similar species that are more tolerant. Hill and Wiegert[40] refer to this as congeneric homotaxis. Toxicant effects are not artificially enhanced by lack of replacement nor are toxicant effects underestimated by lack of competitive pressure.

CRITICISMS OF THE AS-M

Taxonomic Uncertainty

Not many laboratories have experience in taxonomic identification of protozoans. While identification of protozoans to genus requires some skill, requirements are quite comparable to those for working with macroinvertebrate or plankton communities. Biologists working with these groups already have the necessary mind-set and discriminatory skills. Necessary reference materials are more accessible now than previously. Patterson[41] has recently published an illustrated guide that is particularly geared to identification from living materials. The keys in *Ward and Whipple's Freshwater Biology*,[42] a book many aquatic biologists already own, are still useful although the codes of nomenclature have changed.

Specialist literature for greater taxonomic precision is suggested by Patterson.[41] However, there are distinct practical limits to the precision with which entire communities of protozoans can be described. Relying on gross morphology and behavior (a rich source of information lost in preserved samples), some individuals are easily identified to species while others can only be described to family with confidence. Genus is generally the lowest practical taxon for quick identification. Use of definitive techniques (i.e., establishment in pure culture, silver line stain staining, electrophoresis or molecular characterization of rDNA) is impractical and/or impossible for each taxa encountered in 18 complex samples.

While limitation in taxonomic precision is abhorrent to some, toxicity can be identified with the precision available. Because taxonomic distance is inversely related to similarity in sensitivity to toxicants,[27] information may become redundant as one progresses to finer taxonomic scales. Kaesler and colleagues[43] found that generic diversity of macroinvertebrates revealed almost as much about community structure as species diversity in stream surveys. Plafkin and colleagues[4] found that impact assessments based on species-level identifications of macroinvertebrate communities were only slightly different from those based on family-level identifications. Waterhouse and Farrell[44] found high correlations between similarity coefficients based on species-level and genus-level identifications of chironomids.

The hierarchical focus of the AS-M also affects the requirement for taxonomic precision. In the AS-M, a community-level endpoint is monitored (i.e., the numbers of different kinds of organisms maintained). Species are not used as indicators,[45] nor is characterizing the tolerance range or autecology of the species[46] the purpose of the test. Instead, the approach is species neutral. The crucial question is: how many kinds of organisms can be maintained in this environment? To answer this question, the researcher must ask: is this organism the same or different from what I have seen so far? The sequential comparison index,[47] used in evaluating biotic integrity of macroinvertebrate communities in natural systems, is based on this same simple process.

Clearly, many ecological applications require greater taxonomic precision. However, the AS-M has consistently discerned effects of toxic substances using genus-level identifications and a community-level endpoint. The practical limits in taxonomic precision do not overwhelm the discriminatory power of the test.

Sampling Efficiency

Sampling PF substrates by squeezing out as much of the contents as possible is only semiquantitative; some attached species are left behind.[48] Extraction efficiencies of 86±4% have been estimated from volume recoveries. In addition, abundance of individual taxa are generally ranked, not counted.

Toxicant Fate Distortions

While polyurethane itself is stable and inert, it provides a large surface area; this is why it is used as a nonselective sorbent in collecting environmental pollutants for analyses.[49] The PF substrate used in the AS-M can have a total surface area of 0.14 m^2. In addition to adsorbing toxicants, this surface supports organisms that biotransform toxicants. The detritus and sediment that accumulate on substrates as they colonize in natural systems also transform toxicants. Because the AS-M includes a large surface, sediment, detritus, humic material, and bacterial populations involved in biotransformations, interactions between toxicants and these components are included in the test. Key processes such as bioremediation, biomethylation, or mobilization of metals will occur. However, the AS-M is unlikely to model accurately the fate of toxicants in any particular natural system. This is equally true of other toxicity tests, even large ones,[50] and is why the best models of toxicant effects rely on measurements of biologically active toxicants to characterize exposure conditions.

Reports have been made of the inherent toxicity of polyurethane foam,[51] and various causative agents for toxicity have been suggested, such as fatty amine compounds[52] or tin.[49] In order to remove labile toxic substances before testing, we have adopted procedures developed to clean polyurethane for use as a sorbent in analytical applications. Substrates are washed in 1 M nitric or hydrochloric acid, followed by distilled water until the washings are acid free, next with acetone or 95% ethanol, and then thoroughly rinsed and dried.[49]

Variability in Community-Level Endpoints

It has been suggested that emergent endpoints will have lower signal-to-noise ratios than population-level endpoints.[53] However, our data suggest that signal-to-noise ratios of community-level metrics are more than adequate (see

Table 4), and the coefficients of variation associated with selected community-level metrics are quite low. Pratt and Bowers[19] report a mean coefficient of variation for measures of taxonomic richness in control microcosms of 7%. For comparison, mean coefficient of variation for number of young produced by controls in short-term chronic toxicity tests with *Ceriodaphnia* was 18.9% (calculated from Winner[54]). In addition, variation for three to seven single species tests will contribute to the variation of a community-level estimate. Perhaps more important than relative variability is simple relevance. Kelly and Harwell[55] assert that the correspondence of community-level endpoints to valued properties of natural systems that society wishes to protect is sufficient reason to use them regardless of relative sensitivity.

Lack of Standardization in Test Communities

Toxicity tests are generally standardized to ensure repeatability. Given the enormous variability in the structure and function of microbial communities developed at different times or in different places, it is quite reasonable to question the repeatability of tests employing these communities. We have suggested that concentration-response models describing the distribution of tolerances of organisms to toxic substances will be similar across ecosystems as long as a sufficiently large number of species are sampled, and no source community is subjected to selective pressure from contamination in the natural system.[56] While there is no question that different ecosystems respond differently to toxic stress, it may be that toxicant fate distinguishes the systems much more than the proportion of organisms susceptible to increasing levels of stress.

Round-robin tests have not been conducted to evaluate repeatability of the AS-M. However, the effects of toxicants have been evaluated with communities developed at different times, in different systems, and in different laboratories[31] (Niederlehner, unpublished data). Available data suggest reasonable repeatability despite differences in season, temperature, water chemistry, and community composition (Figure 3). The concentrations impairing the taxonomic richness of the community by 20% (IC20s) were estimated for five chemicals (Table 9). No comparisons were made for phenol since only one test had significant reductions in taxonomic richness. Over five toxicants, mean ratio of maximum to minimum IC20 was 2.4. For comparison, Parkhurst and colleagues[32] report a mean maximum/minimum ratio of 3 for intralaboratory variation of IC50s from conventional, short-term, chronic toxicity tests in tests with single chemicals. The individual errors associated with three to seven single species tests would contribute to the precision of a community-level IC20 estimate.[11]

Because selected endpoints in the AS-M seem reasonably repeatable, concern about failure to reconstruct the same community for each microcosm test seems tangential. It may not be necessary to reconstruct the same community for each toxicity test because the underlying tolerance distribution is common

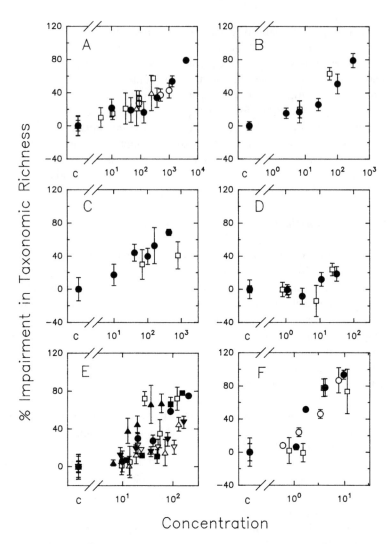

Figure 3   Repeatability of the AS-M. Each unique symbol represents an independent test. (A) chronic zinc toxicity (μg/L), (B) chronic chlorine toxicity (μg TRC/L), (C) chronic ammonia toxicity (μg NH₃-N/L), (D) chronic phenol toxicity (mg/L), (E) chronic copper toxicity (μg/L), and (F) acute cadmium toxicity (mg/L).

to all naive communities. The relevant structural characteristic (i.e., the tolerance distribution) is "standard". Reasonable consistency in community-level response, despite differences in composition, is not unique to the AS-M. Stay and colleagues[57] found that mixed flask microcosms derived from different sources were similar in response to toxic stress. Sheehan[58] found that microcosms derived from different sources consistently ranked the relative hazard of four chemicals.

Table 9   Repeatability of AS-M

	Toxicant				
Parameter	Zn[a,b,c] (μg/L) Chronic	OCl (μg/L) Chronic	NH$_3$ (μg/L) Chronic	Cu[a,b,c] (μg/L) Chronic	Cd[c] (mg/L) Acute
IC20s	21.7	2.7	13.3	23.1	1.2
	31.6	7.1	10.5	30.7	1.3
	61.2			29.4	1.8
	58.8			38.9	
				52.3	
				13.2	
Geometric mean	39.6	4.4	11.8	28.7	1.4
S. D. (log units)	0.22	0.29	0.07	0.20	0.09
C. V. (log units)	13.7	46.0	6.8	14.1	64.1
Min/Max Ratio	2.8	2.6	1.3	4.0	1.5

[a] Communities from different reference systems

[b] Different dosing systems

[c] Different seasons and test temperatures

EXTRAPOLATIONS ACROSS SCALE

The purpose of a toxicity test varies from one application to another. While some tests are comparative (i.e., is this substance more toxic than that?), others are predictive (i.e., what will happen to the natural system if this substance is discharged?). For predictive applications, validity is the ability to predict environmental outcome in natural systems with a degree of precision adequate to management needs. We have suggested that correlations of laboratory toxicity and field impairment are necessary but are not sufficiently rigorous tests of validity for predictive methods.[59] Two measures of toxicity, even those that predict environmental outcome poorly, are likely to be correlated if measured over a wide enough range of concentrations. Instead, the adequacy of predictive toxicity tests must be evaluated based on the size of predictive errors and on the costs of management decisions made on the basis of faulty information. This test of validity demands not only that the measures be correlated but also that predictive errors be within a predefined range, ideally based on management costs but tentatively set at 20% for these analyses.

Predictions of loss of taxonomic richness with increasing acidity based on the AS-M are compared in Figure 4a to existing field data for taxonomic loss across acid gradients. Eilers and colleagues[60] compiled empirical distribution functions describing the tolerance of eight major taxonomic groups to decreasing pH. These tolerance distributions were based on over 1000 observations of minimum pH tolerance in both field and laboratory studies. By interpolating from the graph in the published report, we calculated mean percent loss of taxa over seven taxonomic groups for each 0.5 pH unit between 6.5 and 3.0 (data on sponges were excluded). An additional comparison was made to loss of taxa (equally weighted for major taxonomic groups) in the experimental acidification of Lake 223 from 6.5 to 5.1.[61] The tolerance model based on the AS-M fell

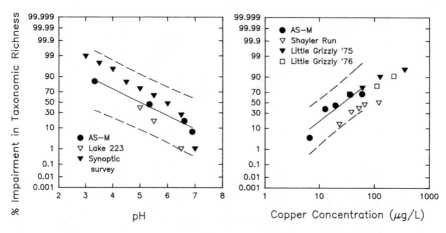

Figure 4   Comparison of tolerance distributions based on AS-M to existing field data across gradients of stress. Dashed lines are the 95% prediction interval for the AS-M based model.

between the two field based distributions. Impairment predicted from the AS-M was correlated to observed impact in the gradient field studies (N=11, p=0.0002, r=0.8945). Further, field observations fell within the 95% prediction interval based on the AS-M. Predictions of environmental outcome based on the AS-M were within 20% of the outcome observed in the synoptic survey at all pH levels. However, the AS-M overestimated impairment in Lake 223 at a pH of 5.5 by >20% (43% vs. 17%). Predictions were within 20% for 10 of 11 cases. The mean error of predictions of impairment compared to both field data sets was 15.1±6.2% (N=11).

In the case of copper (Figure 4b), predicted impairments were correlated to those observed in gradient field studies described by Sheehan and Winner[62] (N=11, p=0.0068, r=0.5758). Predictions of macroinvertebrate response in Little Grizzly Creek based on the AS-M were within 20% of observed outcome. However, predictions from the AS-M consistently overestimated macroinvertebrate response in Shayler Run. Errors ranged from 29% at 23 µg/L (43% predicted vs. 14% observed) to 46% at 120 µg/L (96% predicted vs. 50% observed). Only 6 of 11 predictions were within 20%. The mean predictive error over both systems was 20.5 ± 16.6%.

Given the relevance of speciation and additional chemical influences (e.g., alkalinity, aluminum, calcium, humic materials) to the toxicity of acidity and copper, better correspondence between distributions might be expected if better chemical data were available for each system.

Predictions of environmental effects based on the AS-M were also compared to effects observed contemporaneously in five systems (Table 10) receiving point source discharges. Response at the waste concentration found at the station immediately downstream of the discharge was predicted. Predictions for additional stations would, in most cases, be less accurate due to chemical

Table 10 Description of Validation Studies

Experiment number	Type of system	Type of stress	Type of toxicity test[a]	In stream group[b]	Reference
1	1st order stream	Mixed metals	A	P	66
2	3rd order stream	POTW effluent	ED	M	67
3	3rd order stream	Industrial effluent	ED	M	59
4	3rd order stream	Industrial effluent	ED	M	68, 69
5	stream mesocosms	Selenium	ED	P	5

[a] A = ambient, ED = effluent dilution

[b] P = protozoa, M = macroinvertebrates

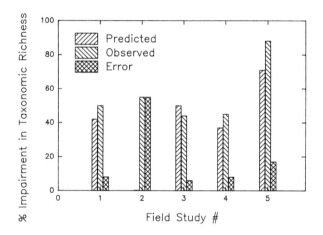

Figure 5 Comparison of predictions of environmental outcome based on the AS-M to contemporaneously observed field impact.

transformations, the added difficulties of predicting subtle rather than gross effects, and station differences in potential for recovery. In four of the five cases, errors were less than the predetermined criterion of 20% (Figure 5). However, in the second case, predictive accuracy was notably poor. Effluent dilution AS-M predicted no impairment at concentrations <30% effluent; ambient AS-M also showed no impairment. However, instream waste concentrations of ~6% resulted in severe losses in the taxonomic richness of macroinvertebrate (55%) and protozoan (46%) communities. Both effluent dilution and ambient *Ceriodaphnia* and fathead minnow short-term chronic tests also failed to predict toxicity at instream waste concentrations. Mean error for predictions of environmental outcome from the AS-M over all five studies was 18.4 ± 21.4%.

While no grand conclusions about the validity of the AS-M are possible from this small and unrepresentative sample of comparisons, we can compare the failure rate and magnitude of errors in predicting environmental outcome from the AS-M to those from other available techniques. Results of *Ceriodaphnia* and fathead minnow toxicity tests were compared to community response in eight streams studied in the U.S. Environmental Protection Agency's (U.S. EPA's) complex effluent toxicity testing program (summarized by Parkhurst and colleagues[32]). Seven predictions were made for only the station immediately downstream of the most toxic discharge; toxicity was not measured at two study sites, but one site was studied over 2 years. Errors varied from 2% to 60%, but the error was <20% in only one of seven comparisons. The mean error was 28.1 ± 17.3%. Based on this summary, the AS-M appears to predict as well or better than other tools available.

CONCLUSIONS

How can small, functionally complex tests contribute to ecotoxicology? Especially in regulatory applications, some questions are efficiently answered by using simple tests. However, because there is great variation in response to toxic stress between organisms, characterizing the range of response of organisms in general is a critical part of predicting environmental response. The AS-M provides an efficient way of estimating distributions of tolerance of organisms. In this way, the relative sensitivity of conventionally tested species can be compared with the much larger number of untested species in a natural community. In addition, the AS-M may provide an independent line of evidence that can increase confidence in more conventional determinations when there is little margin for error. The AS-M can establish cause–effect relationships on endpoints that are directly relevant to management goals but untestable in simpler tests. The AS-M can be a useful scoping tool prior to the great expense and limited design options of large-scale mesocosm tests used to nullify the presumption of risk.

In addition, microcosms, including the AS-M, continue to be an essential heuristic tool. MacArthur[63] states:

> Ecological patterns, about which we construct theories, are only interesting if they are repeated. They may be repeated in space or in time, and they may be repeated from species to species. A pattern which has all of these kinds of repetition is of very general interest because of its generality, and yet these very general events are only seen by ecologists with rather blurred vision. The very sharp-sighted always find discrepancies and are able to say that there is no generality, only a spectrum of special cases.

It is, perhaps, more consistent with the training of a researcher to point out the differences in response to different toxicants or to the same toxicant at different observational scales. However, generalities may also exist and, if confirmed, will allow greater precision in predictive models and better determinations of biotic integrity. We have found reasonable consistency in tolerance distributions for aquatic taxa across spatiotemporal scales and across diversity in community structure and function. We have also found that some community-level endpoints are more sensitive and general than others. In our experience, microcosms on the scale described here are efficient, statistically tractable, repeatable, and they predict relevant environmental outcome as well as other presently available tools.

ACKNOWLEDGMENTS

The authors would like to thank Dick Pratt and Nancy Bowers for providing their raw data from many AS-M tests for our own calculations and comparisons. We would like to thank Darla Donald for editorial assistance. This research was sponsored in part by the Air Force Office of Scientific Research, Air Force Systems Command, U.S. Air Force, under Grant 85–0324. The U.S. government is authorized to reproduce and distribute reprints for governmental purposes, notwithstanding any copyright notation thereon.

REFERENCES

1. Huckabee, J. W., Evaluation of Tests to Predict Chemical Injury to Ecosystems: Microcosms, in *Methods for Estimating Risk of Chemical Injury: Human and Non-human Biota and Ecosystems, SCOPE* , V. B. Vouk, G. C. Butler, D. G. Hoel, and D. B. Peakall, Eds., New York: John Wiley and Sons (1985), pp. 637–674.
2. O'Neill, R. V., D. L. DeAngelis, J. B. Waide, and T. F. H. Allen, *A Hierarchical Concept of Ecosystems,* Princeton, NJ: Princeton University Press (1986).
3. Short-term Methods for Measuring the Chronic Toxicity of Effluents and Receiving Waters to Freshwater Organisms, 3rd Ed., U.S. EPA Report-600/4–91–002 (1991).
4. Plafkin, J. L., M. T. Barbour, K. D. Porter, S. K. Gross, and R. M. Hughes, Rapid Bioassessment Protocols for Use in Streams and Rivers: Benthic Macroinvertebrates and Fish, U.S. EPA Report-444/4–89–001 (1989).
5. Pratt, J. R. and N. J. Bowers, A Microcosm Procedure for Estimating Ecological Effects of Chemicals and Mixtures, *Toxicity Assess.* 5:189–205 (1990).
6. Barnthouse, L. W., G. W. Suter, II, and A. E. Rosen, Risks of Toxic Contaminants to Exploited Fish Populations: Influence of Life History Data Uncertainty and Exploitation Intensity, *Environ. Toxicol. Chem.* 9:297–311 (1990).
7. Odum, E. P., Trends Expected in Stressed Ecosystems, *BioScience* 35:419–422 (1985).
8. Rapport, D. J., H. A. Regier, and T. C. Hutchinson, Ecosystem Behavior Under Stress, *Am. Nat.* 125:617–640 (1985).

9. Schindler, D. W., Detecting Ecosystem Responses to Anthropogenic Stress, *Can. J. Fish. Aquat. Sci.* 44(Suppl. 1):6–25 (1987).

10. Stephan, C. E., D. I. Mount, D. J. Hansen, J. H. Gentile, G. A. Chapman, and W. A. Brungs, Guidelines for Deriving Numerical National Water Quality Criteria for the Protection of Aquatic Organisms and Their Uses, U.S. EPA Report NTIS PB85–227049 (1985).

11. Okkerman, P. C. E., J. V. D. Plassche, W. Slooff, C. J. Van Leeuwen, and J. H. Canton, Ecotoxicological Effects Assessment: A Comparison of Several Extrapolation Procedures, *Ecotoxicol. Environ. Saf.* 21:182–193 (1991).

12. Minns, C. K., J. E. Moore, D. W. Schindler, and M. L. Jones, Assessing the Potential Extent of Damage to Inland Lakes in Eastern Canada Due to Acidic Deposition. III: Predicted Impacts on Species Richness in Seven Groups of Aquatic Biota, *Can. J. Fish. Aquat. Sci.* 47:821–830 (1990).

13. Kleinbaum, D. G. and L. L. Kupper, *Applied Regression Analysis and Other Multivariable Methods,* Boston: Duxbury Press (1978), p. 7.

14. Niederlehner, B. R., J. R. Pratt, A. L. Buikema, Jr., and J. Cairns, Jr., Comparison of Estimates of Hazard Derived at Three Levels of Complexity, in *Community Toxicity Testing, STP 920*, J. Cairns, Jr., Ed., Philadelphia: American Society for Testing and Materials (1986), pp. 30–48.

15. Suter, G. W., II, End points for regional ecological risk assessments, *Environ. Manage.* 14:9–23 (1990).

16. Pomeroy, L. R., The Ocean's Food Web, a Changing Paradigm, *BioScience* 24:499–504 (1974).

17. Laybourn-Perry, J., Protozoan Energetics—session Summary, in *Protozoa and Their Role in Marine Processes*, P. C. Reid, C. M. Turley, P. H. Burkill, Eds., Berlin: Springer-Verlag (1991), pp. 268–279.

18. Curds, C. R. and H. A. Hawkes, *Ecological Aspects of Used-Water Treatment, Volume 1—The Organisms and their Ecology,* Academic Press: London (1975).

19. Pratt, J. R. and N. J. Bowers, Variability of Community Metrics: Detecting Changes in Structure and Function, *Environ. Toxicol. Chem.* 11:451–457 (1992).

20. Giesy, J. P., Multispecies Tests: Research Needs to Assess the Effects of Chemicals on Aquatic Life, in *Aquatic Toxicology and Hazard Assessment: Eighth Symposium, ASTM STP 891*, R. C. Bahner and D. J. Hansen, Eds., Philadelphia: American Society for Testing and Materials (1985), pp. 67–77.

21. Picken, L. E. R., The Structure of Some Protozoan Communities, *J. Ecol.* 25:368–384 (1937).

22. Carpenter, S. R., J. F. Kitchell, and J. R. Hodgson. Cascading Trophic Interactions and Lake Productivity, *BioScience* 35:634–639 (1985).

23. McCormick, P. V. and J. Cairns, Jr., Effects of Micrometazoa on the Protistan Assemblage of a Littoral Food Web, *Freshwater Biol.* 26:111–119 (1991).

24. Matthews, R. A., P. F. Kondratieff, and A. L. Buikema, Jr., A Field Verification of the Use of the Autotrophic Index in Monitoring Stress Effects, *Bull. Environ. Contam. Toxicol.* 25:226–233 (1980).

25. Small, E. B. and J. C. Green, Taxonomy (46-or more-protistan phyla)—Session Summary, in *Protozoa and Their Role in Marine Processes*, P. C. Reid, C. M. Turley, P. H. Burkill, Eds., New York: Springer-Verlag (1991), pp. 1–8.

26. Knoll, A. H., The Early Evolution of Eukaryotes: A Geological Perspective, *Science* 256:622–627 (1992).

27. Mayer, F. L., Jr., C. H. Deans, and A. G. Smith, Intertaxa Correlations for Toxicity to Aquatic Organisms, U.S. EPA 600/x-87/332, Springfield, VA: National Technical Information Service (1987).

28. Slooff, W., J. A. M. Van Oers, and D. De Zwart, Margins of Uncertainty in Ecotoxicological Hazard Assessment, *Environ. Toxicol. Chem.* 5:841–852 (1986).

29. Roesijadi, G., Metallothioneins in Metal Regulation and Toxicity in Aquatic Animals, *Aquat. Toxicol.* 22:81–114 (1992).

30. Gingrich, D. J., D. N. Weber, C. F. Shaw, J. S. Garvey, and D. H. Petering, Characterization of a Highly Negative and Labile Binding Protein Induced in *Euglena gracilis* by Cadmium, *Environ. Health Perspect.* 65:77–85 (1986).

31. Pratt, J. R., N. J. Bowers, and J. M Balczon, A Microcosm using Naturally Derived Microbial Communities: Comparative Ecotoxicology, in *Environmental Toxicology and Risk Assessment, ASTM STP 1179*, W. G. Landis, J. S. Hughes, and M. A. Lewis, Eds., Philadelphia: American Society for Testing and Materials (1993), pp. 178–191.

32. Parkhurst, B. R., W. Warren-Hicks, and L. E. Noel, Performance Characteristics of Effluent Toxicity Tests: Summarization and Evaluation of Data, *Environ. Toxicol. Chem.* 11:771–791 (1992).

33. Connell, J. H. and W. P. Sousa, On the Evidence Needed to Judge Ecological Stability or Persistence, *Am. Nat.* 121:789–822 (1983).

34. Cairns, J., Jr. and B. R. Niederlehner, Adaptation and Resistance of Ecosystems to Stress, AFOSR Report 88–0263 (1991).

35. Stay, F. S., T. E. Flum, L. J. Shannon, and J. D. Yount, An Assessment of the Precision and Accuracy of SAM and MFC Microcosms Exposed to Toxicants, in *Aquatic Toxicology and Hazard Assessment: 12th Volume, STP 1027*, U. M. Cowgill and L. R. Williams, Eds., Philadelphia: American Society for Testing and Materials (1989), pp. 189–203.

36. Hoekstra, T. W., T. F. H. Allen, and C. H. Flather, Implicit Scaling in Ecological Research, *BioScience* 34:148–154 (1991).

37. Pratt, J. R., N. J. Bowers, B. R. Niederlehner, and J. Cairns, Jr., Effects of Atrazine on Freshwater Microbial Communities, *Arch. Environ. Contam. Toxicol.* 17:449–457 (1988).

38. Niederlehner, B. R. and J. Cairns, Jr., Effect of Increasing Acidity on Aquatic Protozoan Communities, *Water Air Soil Pollut.* 52:183–196 (1990).

39. Niemi, G. J., P. DeVore, N. Detenbeck, D. Taylor, A. Lima, J. Pastor, J. D. Yount, and R. J. Naiman, Overview of Case Studies on Recovery of Aquatic Systems from Disturbance, *Environ. Manage.* 14:571–588 (1990).

40. Hill, J., IV and R. G. Wiegert, Microcosms in Ecological Modeling, in *Microcosms in Ecological Research*, J. Giesy, Jr., Ed., DOE Symposium Series 52, NTIS CONF-781101 (1980), pp. 138–163.

41. Patterson, D. J., *Free-Living Freshwater Protozoa: A Color Guide*, Boca Raton, FL: CRC Press (1992).

42. Edmondson, W. T., Ed., *Ward and Whipple's Freshwater Biology*, New York: John Wiley (1959).

43. Kaesler, R. L., E. E. Herricks, and J. S. Crossman, Use of Indices of Diversity and Hierarchical Diversity in Stream Surveys, in *Biological Data in Water Pollution Assessment; Quantitative and Statistical Analyses, STP 652*, K. L. Dickson, J. Cairns, Jr., and R. J. Livingston, Eds., Philadelphia: American Society for Testing and Materials (1978), pp. 92–112.

44. Waterhouse, J. C. and M. P. Farrell, Identifying Pollution Related Changes in Chironomic Communities as a Function of Taxonomic Rank, *Can. J. Fish. Aquat. Sci.* 42:406–413 (1985).

45. Foissner, W., Taxonomic and Nomenclatural Revision of Sladecek's List of Ciliates (Protozoa:Ciliophora) as Indicators of Water Quality, *Hydrobiologia* 166:1–64 (1988).

46. Corliss, J. O., The Interface Between Taxonomy and Ecology in Modern Studies on the Protists, *Acta Protozool.* 31:1–9 (1992).

47. Cairns, J., Jr., D. Albaugh, F. Busey, and M. Chanay, The Sequential Comparison Index — A Simplified Method for Nonbiologists to Estimate Differences in Biological Diversity in Stream Pollution Studies, *J. Water Pollut. Control Fed.* 40:1607–1613 (1968).

48. Bamforth, S. S., The Variety of Artificial Substrates used for Microfauna, in *Artificial Substrates*, J. Cairns, Jr., Ed., Ann Arbor, MI: Ann Arbor Science (1982), pp. 115–130.

49. Braun, T., J. D. Navratil, and A. B. Farag, *Polyurethane Foam Sorbents in Separation Science,* Boca Raton, FL: CRC Press, Inc. (1985).

50. Kaushik, N., K. Solomon, G. Stephenson, and K. Day, Use of Limnocorrals in Evaluating the Effects of Pesticides on Zooplankton Communities, in *Community Toxicity Testing, ASTM STP 920*, J. Cairns, Jr., Ed., Philadelphia: American Society for Testing and Materials (1986), pp. 269–290.

51. Schafer, P. M. and M. Moser, Sterilization Method Determines Effect of Toxic Foam Plug on Algal Cultures, *Water Res.* 23:1323–1326 (1989).

52. Bach, J. A., R. J. Wnuk, and D. G. Martin, Inhibition of Microbial Growth by Fatty Amine Catalysts from Polyurethane Foam Test Tube Plugs, *Appl. Microbiol.* 29:615–620 (1975).

53. Kooyman, S. A. L. M., Toxicity at the Population Level, in *Multispecies Toxicity Testing*, J. Cairns, Jr., Ed., New York: Pergamon Press (1985), pp. 143–164.

54. Winner, R. W., Evaluation of the Relative Sensitivities of 7-d *Daphnia magna* and *Ceriodaphnia dubia* Toxicity Tests for Cadmium and Sodium Pentachlorophenate, *Environ. Toxicol. Chem.* 7:153–159 (1988).

55. Kelly, J. R. and M. A. Harwell, Indicators of Ecosystem Response and Recovery, in *Ecotoxicology: Problems and Approaches*, S. A. Levin, M. A. Harwell, J. R. Kelly, and K. D. Kimball, Eds., New York: Springer-Verlag (1989), pp. 9–40.

56. Cairns, J. Jr., B. R. Niederlehner, and J. R. Pratt, Evaluation of Joint Toxicity of Chlorine and Ammonia to Aquatic Communities, *Aquat. Toxicol.* 16:87–100 (1990).

57. Stay, F. S., A. Katko, C. M. Rohm, M. A. Fix, and D. P. Larsen, The Effects of Atrazine on Microcosms Developed from Four Natural Plankton Communities, *Arch. Environ. Contam. Toxicol.* 18: 866–875 (1989).

58. Sheehan, P. J., Statistical and Nonstatistical Considerations in Quantifying Pollutant-Induced Changes in Microcosms, in *Aquatic Toxicology and Hazard Assessment, Twelfth Volume, ASTM STP 1027*, U. Cowgill and L. Williams, Eds., Philadelphia: American Society for Testing and Materials (1989), pp. 178–188.

59. Niederlehner, B. R., K. W. Pontasch, J. R. Pratt, and J. Cairns, Jr., Field Evaluation of Predictions of Environmental Effects from a Multispecies-Microcosm Test, *Arch. Environ. Contam. Toxicol.* 19:62–71 (1990).

60. Eilers, J. M., G. J. Lien, and R. G. Berg, Aquatic Organisms in Acidic Environments: A Literature Review, Technical Bulletin No. 150, Madison, WI: Department of Natural Resources (1984).

61. Schindler, D. W., S. E. M. Dasian, and R. H. Hesslein, Losses of Biota from American Aquatic Communities Due to Acid Rain, *Environ. Monit. Assess.* 12:269–285 (1989).

62. Sheehan P. J. and R. W. Winner, Comparison of Gradient Studies in Heavy-Metal-Polluted Streams, in *Effects of Pollutants at the Ecosystem Level*, P. J. Sheehan, D. R. Miller, G. C. Butler, and P. Bourdeau, Eds., New York: John Wiley and Sons (1984), pp. 255–271.

63. MacArthur, R. L., The Theory of the Niche, in *Population Biology and Evolution*, R. C. Lewontin, Ed., Syracuse, NY: Syracuse University Press (1968) pp. 159–176.

64. Cairns, J., Jr., J. R. Pratt, B. R. Niederlehner, and N. Bowers, Structural and Functional Responses to Perturbation in Aquatic Ecosystems, Final Technical Report AFOSR 85–0324, Air Force Office of Scientific Research, Springfield, VA: NTIS AD-192071/9/GAR (1988).

65. Torsvik, V., J. Goksoyr, and F. L. Daae, High Diversity in DNA of Soil Bacteria, *Appl. Environ. Microbiol.* 56:782–787 (1990).

66. Shen, Y., A. L. Buikema, Jr., W. H. Yongue, Jr., J. R. Pratt, and J. Cairns, Jr., Use of protozoan communities to predict environmental effects of pollutants, *J. Protozool.* 33:146–151 (1986).

67. Pratt, J. R., J. Mitchell, R. Ayers, and J. Cairns, Jr., Comparison of Estimates of Effects of a Complex Effluent at Differing Levels of Biological Organization, in *Aquatic Toxicology and Environmental Fate: Eleventh Volume, ASTM STP 1007*, G. W. Suter, II, and M. A. Lewis, Eds., Philadelphia, PA: American Society for Testing and Materials (1989), pp. 174–188.

68. Pratt, J. R., N. B. Pratt, and B. R. Niederlehner, Microbial Community Toxicity Testing, Report to the Research Triangle Institute, Research Triangle Park, NC (1985).

69. Brown, B., Tennessee Department of Environment and Conservation, Unpublished results (1992).

CHAPTER **9**

The Use of Constructed or Artificial Ponds in Simulated Field Studies

James H. Kennedy, Philip C. Johnson, and Zane B. Johnson

INTRODUCTION

There is an increased worldwide interest in the use of surrogate ecosystems for the evaluation of the fate and effects of contaminants in aquatic ecosystems, as indicated by the number of recent symposia[1-3] and workshops[4-7] held in North America and Europe. Surrogate ecosystems are increasingly utilized as experimental and regulatory tools to study the fate and effects of xenobiotics in aquatic ecosystems.[8] For example, the Federal Insecticide, Fungicide and Rodenticide Act (FIFRA) requires an assessment of ecological risk of pesticides before a pesticide can be granted registration in the U.S. Since the mid-1980s the U.S. Environmental Protection Agency (U.S. EPA) has adopted the use of freshwater mesocosms (experimental ponds) as surrogate natural systems in which ecosystem level effects of pesticides can be measured for use in the ecological risk assessment process. "Risk" includes both toxicity and exposure elements. Outdoor microcosms and mesocosms are particularly effective in this regard as they involve the exposure of pesticides to complex assemblages of organisms in the aquatic environment.[5]

Mesocosms can be defined as either physical enclosures of a portion of a natural ecosystem or outdoor manmade structures such as ponds or stream channels that (closely) simulate the natural environment.[9] Constructed systems, additionally, are effective tools in aquatic research because they (are simplified enough to) act as surrogates for important cause/effect pathways in natural systems,[9,10] yet retain a high degree of environmental realism relative to laboratory single species bioassays.[10] Voshell[1] specifies that the size and complexity of mesocosms is sufficient for them to be self-sustaining, making them suitable for long-term studies. While these terms describe many properties of mesocosms, they place no limitations on their size. Increasingly, researchers have divided surrogate ecosystems on the basis of size into two categories; microcosms and mesocosms. Microcosms can be defined, following Giesy and Odum[11] as "...artificially bounded subsets of naturally occurring environments which are replicable, contain several trophic levels, and exhibit system-level properties."

Many researchers have viewed microcosm studies as small-scale indoor systems. The term microcosm, however, has recently gained widespread use, especially in North America, to include outdoor test systems significantly smaller than mesocosms but still large enough to incorporate many or all of the same ecological components as mesocosms.[6] Cairns[10] does not distinguish between mesocosms and microcosms because both encompass higher levels of biological organization and have high degrees of environmental realism. Historically, a wide size range of outdoor surrogate ecosystems, here referred to collectively as mesocosms, have been used to study ecological interactions or ecotoxicological responses of communities. In order to avoid confusion in the future, the organizing committee of the European Workshop on Freshwater Field Tests (EWOFFT) delineated surrogate ecosystems on the basis of size, with microcosms defined as systems that contained less than 15 m^3 of water volume and mesocosms for test containers of 15 m^3 or larger volumes of water. The EWOFFT characterized experimental stream channels on the basis of length, defining microcosms as shorter and mesocosms as longer than 15 meters.

An array of experimental test systems have been used to study the fate and effect of pesticides and other contaminants in aquatic ecosystems. Microcosm systems have included tanks of various construction (earthen, concrete, fiberglass, stainless steel, or small enclosures). Mesocosm test systems include earthen ponds or enclosures that isolate a section of a larger body of water. These latter systems are commonly referred to as limnocorrals if they are not associated with the shoreline and littoral corrals if they enclose aquatic and shoreline areas. There have been a number of recent reviews of mesocosm and outdoor microcosm research,[4,8,12] and another would be superfluous. The present paper concentrates on critical issues of outdoor surrogate ecosystem experimental design that need to be considered and analytical techniques to enhance the ability of these systems to resolve ecosystem responses. Topics discussed are illustrated with examples of studies conducted using either outdoor microcosms or mesocosms.

SCALE

Factors related to scale are a major design consideration that may effect the assessment of ecotoxicological effects of pesticides in aquatic ecosystems, and scale will influence the cost of studies.[7] The selection of scale is critical in the design of aquatic ecosystem studies and can influence the biological effects that will be observed when systems are exposed to various stressors. The scales primarily addressed in research are spatial and temporal and, though often treated separately in ecological study, are more appropriately integrated because they occur concurrently in natural systems.[13] Spatial scales encompass the physical range organisms may inhabit or the space in which physical, chemical, and geological processes occur, while the temporal scale pertains to time as relating to an individual's or population's activities, life spans, or even the time required for key interactions or processes. The temporal aspect additionally includes natural periodicities, whether diel, annual, or successional, in which such events occur. Frost et al.[14] state that practical considerations rather than appropriate choices of scale have constrained lake research and that scale has typically not been incorporated into sampling protocols and experimental designs.

Research goals should be the primary determinant of the choice of scale; with relevant biological and physicochemical parameters guiding this process. Sampling regimes should be compatible with these scales. In an oversimplified example, it would be unthinkable to examine the effects of a given manipulation on the diel vertical migrations of zooplankton species by sampling weekly or by sampling only the littoral zone. Sampling schedules and techniques compatible with the appropriate scales of the study will maximize the information content of individual samples and perhaps even reduce the number of sampling replicates required to identify effects.

SCALING CONSIDERATIONS IN SURROGATE SYSTEMS

Critical physical and chemical processes behave differently as both a function of scale and perhaps as a contributor to scaling differences. Giddings and Eddlemon[15] reported microcosm size as having a contributing effect to arsenic loss rates from water. Differences in chemical fate were attributed to differential surface area:volume ratios between the two sizes of systems used. In this study, higher arsenic concentrations remained in the water column of larger microcosms, allowing greater amounts of arsenic to be taken up by algae.[15] It is therefore apparent that such scaling considerations can affect chemical exposure and may have implications for population and community level structure and resultant functional attributes.

The influences of scale on community structure were addressed by Stephenson et al.,[16] who reported increased complexity and variability in the spatial distribution of freshwater zooplankton with increased limnocorral sizes, while "edge effects" observed in the smallest enclosures resulted in more homogeneous planktonic communities. No variables other than enclosure dimensions were manipulated in this study; no perturbations were introduced, yet important differences in planktonic community were observed. It could be inferred then, that when further manipulated, the responses of systems of differing size and complexity might be increasingly divergent.

In experiments introducing methoxychlor to enclosures of three sizes, Solomon et al.[17] reported that reductions in zooplankton abundance in the smallest enclosures were numerically less, and their recovery rates were faster relative to the larger enclosures. Dissipation of methoxychlor from the water column differed among the three enclosure sizes, occurring most rapidly in the smallest enclosures. This is perhaps attributable to the periphytic community associated with the greater surface area to volume ratio in the smallest systems.[17] Dudzik et al.[18] similarly cite the prevalence of biological and chemical activity on the sides and bottoms of microcosms as an important consideration in utilizing such smaller scale systems. Lundgren[8] reviewed the effects of size in marine and freshwater artificial ecosystem research.

Outdoor concrete microcosms (1.9 m^3) were used to study the chemical fate and effects of cyfluthrin, a pyrethroid insecticide.[19,20] Results from microcosms were compared with a concurrent cyfluthrin pond mesocosm 635 m^3 study. Pesticide loadings were scaled by system volume, with the same nominal concentrations in mesocosms and microcosms. The experimental design and number of replicates were held constant in both systems. Aqueous cyfluthrin concentrations (sampled 1 h after application) were generally higher in microcosms, while aqueous half-life was shorter (more rapid dissipation) in the smaller systems. Sediment residues of cyfluthrin were higher in the microcosms, compared with mesocosms, for a given pesticide exposure level. Howick et al.[21] noted similar effects of scale for an organophosphate pesticide (sulprofos) tested in outdoor fiberglass microcosms and pond mesocosms. They also observed higher pesticide residues in sediments within the smaller systems.

VARIABILITY

It has been a long accepted tenet in ecology that ecosystems exhibit inherent variation in the forms of structural heterogeneity and resultant functional dynamics. This "background" variation, which occurs throughout a given system and among (similar) systems, consists of both spatial and temporal components, which must be considered in ecological research. The scale of these aspects is also relevant to research efforts; the temporal component, for example, might be considered on either a diel, seasonal or annual basis, or through ecological time as pertaining to succession events. The failure to observe patterns (predicted or otherwise) in ecotoxicological experimental manipulations is too often a result of the spatiotemporal scales selected.

Environmental heterogeneity, and its implications for species distributions, dispersion patterns, and interactions, has prompted ecologists to develop and utilize surrogate systems, whether laboratory microcosms, *in situ* enclosures, or artificial outdoor "units", to reduce variability and enhance predictability in their research. This is especially true if the study involves the manipulation of either biotic or abiotic components of the system. In field experiments, one variable is manipulated while others vary naturally, but it may not be possible to determine if a nonmanipulated variable has either altered or masked an expected result.[22] Community structure and spatial heterogeneity can be controlled to a greater extent in artificial (constructed) systems relative to natural ones. This makes these systems logistically more manageable and more replicable for statistical analyses. In appreciably heterogeneous systems, even small alterations of an independent variable may result in unpredicted, sweeping changes in community structure and function that are difficult to quantify and, perhaps more importantly, reproduce in subsequent studies. Continuous colonization of outdoor systems tend to cause them to follow their own trajectory through time.[3,12] Consequently, even control systems tend to diverge from one another, increasing among system variability.

The use of artificial aquatic systems to study community structure (i.e., what determines it and what affects it) serves a two-fold purpose. Initially, it allows the ecologist to address ecological hypotheses on a manageable scale, with replication and with controls. (Both replication and adequate control conditions are difficult to produce in natural experiments.) Secondly, these systems provide the ecotoxicologist with models of how ecosystems actually function [in the absence of perturbation] so that the effects of toxins in ecosystems might be better separated from natural events such as succession or from inherent variation. Crow and Taub[23] emphasized the importance of this latter aspect when predicting pollutant effects in ecosystems.

The confounding influence of system variability in ecotoxicological studies involving constructed systems has long been recognized,[18] but no uniform approach to a solution has been reached. Giddings and Eddlemon[15] assessed the microcosm variability for the purpose of evaluating the effect of size on stability of selected parameters, such as pH, dissolved oxygen (DO), conductivity, and gross species composition. Other researchers[24] have assessed inherent system variability and determined the amount of sampling replication required to detect treatment effects. Given that variability is inherent in all biological systems, the limiting of variability (whether natural or experimenter induced) is critical to our ability to detect "true" ecological effects. Variability may occur either within or among experimental systems. Replication of treatments and the use of controls are necessary to help distinguish natural variation from the effects of treatment.[25,26]

Sampling replication is necessary to assess within-system heterogeneity in model ecosystems. Assessing variability through such methods as examination of coefficients of variation[27] and determining the number of sampling replicates that would be adequate to ensure representative sampling becomes critical in ecological research. Studies of stream benthos, however, indicate the number of samples required to obtain adequate representation of the community would be quite high and no doubt impractical.[27–29] It is probable that commonly employed sampling regimes in lentic and marine research may similarly underestimate inherent variability in these habitats. Pratt and Bowers[30] point out that, although the number of replicates needed to detect effects of a given magnitude can be determined *a priori* by using estimates of variability, such estimates often do not match the availability of research resources (personnel, space, or time). Therefore, sampling must focus on those variables that convey scientific meaning and provide investigators with resolving power adequate to identify differences. At the present time these variables are primarily structural,[31] however, similar functional assessments should be evaluated and employed when possible.

REDUCING SYSTEM VARIABILITY

One traditional alternative for reducing within-system variation has been to decrease the complexity of surrogate ecosystems. In some cases, this approach may result in simplification to the extent that model systems will not resemble the natural systems they are attempting to "mimic", thereby affecting predictability[11] and applicability. However, mimicking natural systems may not provide the best experimental models according to Maciorowski,[32] who further states that the challenge in ecotoxicological research is to find those phenomena that *can be* simplified to several salient interactions.

Methods of limiting among-system variability sometimes employ engineering design features and physical manipulations of organisms. One routinely applied method in mesocosm, and sometimes in microcosm, research involves the circulation of water among the systems prior to commencement of the study.[33–38] Systematic "seeding" of the systems with biota and sediments from mature ponds may minimize variability resulting from nonuniform distributions of macroinvertebrates and macrophyte seeds.[39]

Variability is not a characteristic that can be easily isolated. It is inherently integrated with other aspects of both surrogate and natural systems. System maturity affects the degree of variability; therefore, it is necessary for systems to have adequate acclimation periods that are dependent on both system size and study goals. Equilibration periods within constructed systems have been employed to enhance system stability, to increase similarity among systems, and to provide greater system realism. Systems need not simulate natural conditions exactly at all levels to be ecologically analogous to them, and achieving "dynamic equilibrium" may not be necessary (or possible), but adequate time is required to establish a number of interacting functional groups.[11] Due to seasonal succession patterns, Giesy and Odum[11] point out that studies conducted in nonequilibrium systems may be more useful than ones carried out under equilibrium conditions.

ROLE OF FISH IN EXPERIMENTAL DESIGN

Current U.S. EPA guidelines[40] require that mesocosm studies include a reproducing population of bluegill sunfish (*Lepomis macrochirus* Rafinesque).

Presumably these fish and their offspring are integrators of system-level processes and differences in numbers, biomass, and size distribution between exposure levels provide requisite endpoints for risk management decisions.[40] From a regulatory perspective, evaluation of fish survival, growth, and reproduction is among the most important endpoints of a mesocosm study.

It is well established that fish populations have both direct and indirect effects on ecosystem function. In littoral enclosures, Arumugam and Geddes[41] determined that larval golden perch (*Macquaria ambigua*) predation significantly affected the densities of many zooplankton species and the overall zooplanktonic community structure. Survival of fish among the five enclosures that were stocked with larval golden perch was highly variable (15.5 to 47%) but comparable to that of the pond in which their enclosures were placed. Cluster analysis of zooplankton community similarity values indicated that enclosures that had experienced the highest fish mortality were more similar to fishless "controls" than enclosures with lower mortality, which clustered with the surrounding pond zooplankton communities.[41]

Fish predation is known to alter plankton community composition,[42–44] and the presence of fish in limnocorral or microcosm experiments may alter nutrient dynamics and cycling.[45,46] For example, during an outdoor microcosm experiment, Vinyard and others[44] found that filter feeding cichlids altered the "quality" of nitrogen (shifting dominant form) and decreased limnetic phosphorus levels via sedimentation of fecal pellets. Additionally, unequal fish mortality among replicate microcosms may influence nutrient levels independently of any other treatment manipulations.[47]

In separate limnocorral studies, Brabrand et al.[48] and Langeland et al.[49] similarly concluded that fish predation alters planktonic communities in eutrophic lakes and that the very presence of certain fish species may contribute to the eutrophication process. Both of these studies offered a number of interesting hypotheses regarding fish effects in limnetic systems, unfortunately, the experimental designs of these studies lacked treatment replication, limiting their inferential capability.

Chemical registration studies by Hill et al.,[50,51] Howick et al.,[39] Johnson et al.,[20] Morris et al.,[52] and Mayasich et al.[53] have determined that the abundance of young bluegill in mesocosm experiments obscured or complicated the evaluation of pesticide impacts on many invertebrate populations. This is consistent with Giesy and Odum's[11] suggestion that higher trophic levels assert a controlling influence on lower trophic levels in microcosms being used for effects studies. Ecological research with freshwater plankton and pelagic fish communities indicates that both "top-down" and "bottom-up" influences affect planktonic community structure and biomass.[47,54–56] These relationships have not been investigated to the same degree in littoral zone communities, and the role of benthic macroinvertebrates in these trophic relationships requires further study. Along these lines, Deutsch et al.[57] stocked largemouth bass in their pond mesocosms in order to control unchecked bluegill population growth, thereby potentially limiting among-system variability and providing a more natural surrogate system. However, the desirability of adding bass to mesocosms must be balanced against possible increases in experimental error variances that may result from differential predation on bluegill if variable bass mortality occurs in the ponds.[58] The only way to control variability in predation of bluegill would be to maintain equal levels of predator mortality in all ponds.

The requirement of using a single fish test species (bluegill sunfish) in mesocosm experiments may not be sufficiently protective of natural fish communities for a number of reasons. First, the inherent sensitivity of other fishes compared with bluegill is not known with any degree of certainty. Second, due to a variety of life history adaptations, other fish might experience differential exposure to chemicals. Surface dwelling fish, such as top-minnows, would potentially be exposed to high initial pesticide concentrations found in the surface layer following treatment. Contaminants that sorb to sediments (including many pesticides) might be expected to impact bottom-feeding fish selectively. Drenner and others[59] studied the effects of a pyrethroid insecticide on gizzard shad, *Dorosoma cepedianum*, in outdoor microcosms. These fish are filter feeders and commonly have large amounts of bottom sediments and detritus in their digestive systems. This study[59] is unique in using a "nonstandard" fish species. Similar field studies utilizing other fish species should be pursued in order to evaluate the influence of feeding behavior and habitat selection on chemical exposure. Following appropriate research, it is conceivable that a multispecies assemblage (i.e., surface feeder, water-column planktivore, and bottom feeder) might eventually be used to represent better potential impacts to natural fish communities.

Future research efforts should determine whether more complex fish communities containing additional trophic levels could be used and whether that would enhance our ability to predict ecological impacts in natural aquatic ecosystems. Criteria for fish stocking levels, regardless of species, are highly dependent on system size. Therefore, the numbers, biomass, and potential for reproduction of the fish stocked should be selected accordingly.

ANALYSIS OF MESOCOSM DATA

INFERENTIAL STATISTICAL METHODS

During the course of a mesocosm study, volumes of data are collected on a wide spectrum of biological and physicochemical endpoints. While direct and indirect effects are reported and discussed in all of these studies, the majority of analyses have relied on changes in population density (total, subgroup, or individual) to detect and measure effects. Regression, t-test, analysis of variance, and multiple range tests are the most commonly used statistical tests to determine the significance of treatment effects.[37]

Current analytical approaches to ecosystem-level testing, particularly in the regulatory arena, have generally employed traditional methods (ignoring potentially important new techniques). A recent discussion of the research applications of model aquatic systems[37] states that ANOVA procedures with multiple range tests are the most widely used statistical procedures in such research. This approach is consistent with U.S. EPA requirements for pesticide registration purposes and hypothesis testing in general.[58] These inferential analyses are often combined with descriptive statistics such as diversity indices and coefficients of variation.[60] Additionally, studies may use more than one type of inferential statistic when examining different effects[61] or for purposes of evaluating the relative strengths and applicabilities of the statistical procedures.[62] While these approaches may be appropriate for current regulatory requirements, innovative approaches in experimental design and data analysis may be necessary if we are to advance the science of ecotoxicology.

COMMUNITY LEVEL ANALYSIS

Diversity indices have been used extensively in the past for combining information about the community into a single number. Diversity is composed of two distinct measures; species richness (the total number of species) and taxonomic evenness (how abundance data are distributed among the species).[63] Research has shown that diversity indices are not as sensitive an indicator of community *structure* as simpler measures such as species richness.[64] Diversity indices assume that when aquatic communities are stressed, the number of species and taxa evenness are reduced, resulting in a lowered index value. However, diversity sometimes does not decrease in stressed systems or even increases due to increased evenness.[63] A sensitivity study by Boyle et al.[65] demonstrated that diversity indices tend to be insensitive, unstable, and inconsistent in their response to stressors.

Similarity indices offer a method of measuring changes in taxonomic composition in stressed ecosystems. Until recently, however, their utility has been limited by a lack of statistical methods to compare multiple similarity indices. In the past, similarity indices compared community structure by treatment level or sampling site, based on means of the replicates, which were then subjectively evaluated by the investigator. A recent method, SIGTREE, developed by Nemec and Brinkhurst[66] provides an inferential procedure for testing hypotheses concerning changes in community taxonomic composition. This nonparametric statistical method uses the Percent Similarity (PS) index discussed by Bray and Curtis[67] and a bootstrap technique to identify statistically significant clusters.[66,68,69] The bootstrap technique determines the statistical significance by iterative replacement and subsequent evaluation. Essentially, empirical data are used to establish the probability distribution used for calculating significance. This analysis provides researchers an avenue for examining direct and indirect effects and responses in a more objective and integrated manner, which both strictly descriptive and single-factor inferential analyses have not allowed. The basic idea of the bootstrap involves determining the variability in an unknown distribution from which your data are drawn by resampling points from replicate samples, with replacement, to create a series of bootstrap samples of the same size as the original data. Each of these "bootstrapped samples" is analyzed using Percent Similarity, and the variation among the resulting estimates indicates the size of error involved in making estimates from the original data.[70]

Ecotoxicological studies should examine indirect responses of the system being studied in more detail and explore the possibility of including more integrative analyses than have been used. We need to continue to explore methods such as community similarity particularly when "bootstrap" or "jack-knife" methods allow estimation of associated probability distributions.

Other potentially useful analytical methods range from simple graphical representations to ecological modeling approaches such as stability analysis, pattern recognition, path analysis, and energy flow models. These evaluative tools may eventually be used to integrate structural and functional responses and provide a more holistic measure of change. However, before these methods can be used in a regulatory framework, researchers need a better understanding of how test systems function and require much more information concerning the relative usefulness and sensitivity of these mathematical applications. As new analyses are adopted, users need to keep in mind that the inclusion of any one of them in the data analysis does not preclude the use of sound ecological judgment in their interpretation of the results.[71]

MULTIPARAMETER APPROACHES

Multiparameter indices may provide another approach for evaluating mesocosm data. These methods are patterned after the Index of Biotic Integrity (IBI) developed for midwestern streams.[72,73] The U.S. EPA has developed Rapid Bioassessment Protocols (RBPs) for fish and macroinvertebrates[74] in wadeable streams. Similar approaches for large rivers, lakes, wetlands and estuaries are either under development or planned for the future. These methods have been used for point source and nonpoint source field studies, and are the basis for development of numeric biological criteria. Field methods developed for RBPs (such as kick net sampling) may not be appropriate for use in pesticide regulatory studies, for instance, however the concept of combining a number of metrics (e.g., taxonomic richness, number of Ephemeroptera/Trichoptera/Plecoptera, Hilsenhoff Biotic Index, percent shredders) into a single value for decisionmaking may be a valuable way to integrate many separate endpoints.

Fish diversity is typically low or nonexistent in mesocosm and microcosm studies, however macroinvertebrates, zooplankton and algal indices might work quite well. Indeed, applying these techniques to artificial ponds or tanks would be straight forward since the "reference condition" would be determined from the control treatment. More research will be needed before appropriate, sensitive endpoints are determined for pond systems. Before we embrace new measures of ecosystem responses as regulatory requirements, we need to understand thoroughly the assumptions and limitations of these tests. In addition, we need to comprehend clearly what the endpoints of these tests are telling us. Whatever methods are ultimately adopted for measuring chemical stressors and ecological interactions, if they are to be used for regulatory purposes, it will be important that they have probability attached to them so that we can extrapolate with definable error.

VALIDATION FOR MESOCOSMS

The future of model ecosystems in predicting biological effects is likely to be contingent upon their ability to simulate accurately critical higher-level processes inherent in natural systems. Cairns[75,76] and Smith[77] have repeatedly called for validation of ecotoxicological results. Until the degree of ecological validity can be established, the predictions and models generated from microcosm and mesocosm research will be of limited inferential value in both risk assessment and ecological research. The issue of scale, discussed earlier, will no doubt continue to complicate this validation process, as critical system aspects will potentially differ with system size.

No matter what techniques are adopted for analysis of ecosystem data, we need to understand how the structural and functional components of the test systems relate to natural systems, in order to provide risk assessments of pesticide impacts to natural ecosystems using data collected from surrogate manmade ponds. Our concerns for protection of ecosystems should, in reality, focus on a regional, continental, or global scale, not on a small lake or stream segment.[78]

Concerns about field validation of laboratory toxicity testing results have been widespread and represent legitimate concerns.[10,76,77] Many of these same criticisms can be applied to microcosm and mesocosm tests. It is largely assumed that responses (chemical fate and ecological effects) within these systems parallel effects found in larger natural ecosystems. This important supposition has rarely been tested, however. Three potential approaches for validating surrogate systems are as follows:

1. Evaluate toxicant impacts in whole-lake ecosystems and mesocosms (or microcosms) simultaneously in order to verify that these different scales behave in a similar manner.

   This direct approach would yield the most straightforward comparisons, but lack of replication in whole-lake systems would limit the inferential capability necessary to draw conclusions.[79] Furthermore, the number of whole-lake systems available for systematic toxicological manipulation is relatively small, and wide-spread validation might prove both expensive and difficult or even undesirable.

One example of this process is provided by Schindler,[31] who compared limnocorral mesocosms to whole-lake acidification and eutrophication studies in the Experimental Lakes Area (ELA) in Ontario, Canada. This research established general agreement between these two scales for chemical and plankton endpoints, if sediment-water interfaces were left intact and experiments were limited to relatively short durations. Mesocosms containing fish were more problematic. As our earlier discussion suggests, the role of fish loadings and community composition deserves more detailed study.

Schindler's excellent work is an important first step in the validation process. Similar efforts would enhance our knowledge of how closely microcosms and mesocosms parallel larger scale systems.

2. Formulate predictions of mesocosm system behavior (i.e., model effects before field testing), and compare results with *a priori* assumptions. As Cairns[10] has stated:

If ecologists can learn to construct mesocosms that replicate important characteristics of natural ecosystems, this will provide unmistakable evidence that they really understand how these ecosystems function.

This approach may eventually become feasible as the behavior of ecological systems is more fully understood, but current state-of-the-science is not adequate to fully predict mesocosm structural and functional changes.

3. Finally, one might submit that "if my model system has sufficiently similar components and functional properties as the natural systems it is simulating, toxicological effects should also be similar." This approach involves more extrapolation than the first approach, as not all of the natural system components can be adquately incorporated, but also might be easier to accomplish and less expensive. As such, preliminary evaluation of mesocosm "realism" might constitute an initial (yet essential) first step in the validation process.

If this approach is pursued, criteria for judgment might include similar biotic composition, comparable population densities (particularly fish loadings), and parallel functional attributes between the natural and surrogate systems.

An additional issue to consider during the design of mesocosm experiments and subsequent validation is how "generic" a surrogate system should be. Highly generic experimental systems might prove difficult to validate in the field, whereas more site-specific designs may prove more tractable since less extrapolation would be involved.

It would seem, intuitively, that results from mesocosm and microcosm tests would more closely approximate effects observed in natural systems compared with laboratory bioassays on individual species. Until these linkages are explored and understood, however, a degree of uncertainty will exist regarding the strength of this connection.

Ultimately, the actual value in mesocosm testing for the risk assessment of pesticides is not the ability to measure effects in surrogate ecosystems but rather their potential to predict responses in natural ecosystems. Before these types of extrapolations can be made we must evaluate the validity of mesocosms to predict ecosystem level responses. Modeling approaches, such as those discussed in this chapter, may ultimately prove to be useful in extrapolating from mesocosms to natural ecosystems.

The mesocosm studies being performed under the FIFRA statute are measuring hundreds of biotic and abiotic aquatic parameters and are producing voluminous data sets. We have an inadequate understanding, however, of the biological processes that occur in many ecosystems.[80] The information needed to *begin* answering questions concerning system function, ecological interactions among species, and the development and refinement of ecosystem level evaluation tools, are contained in these existing data sets. A systematic synthesis of the existing information will aid mesocosm testing in reaching its full potential as a tool for prediction of impacts. Unfortunately, time constraints imposed for completion of these resource-intensive field studies and the proprietary nature of these data have left an untapped source of information concerning ecosystem responses. Ecologists, toxicologists, and the chemical industry should collaborate to evaluate this wealth of data from the standpoint of ecosystem response. The most useful information may ultimately come about from innovative procedures that cross traditional disciplinary boundaries[32] and lead to a more defensible regulatory tool that incorporates ecosystem level analyses in the ecological risk assessment process. It is unlikely that the analysis of mesocosm data will rely on any single analytical method as a panacea for the risk assessment of agricultural chemicals. Rather, these decisions will involve a synthesis of techniques that embrace biological concepts, multivariate statistics, modeling and experimental approaches.

SUMMARY

In this chapter, we have discussed the role of constructed artificial ponds and microcosms as model ecosystems for use in simulated field studies. Critical design factors to consider include the role of temporal and physical scale, natural variability between and within replicates, and consideration of fish loadings and fish community structure. This chapter also reviewed the status of analytical tools necessary for interpretation of mesocosm results and discussed assessment methods that should be considered in the future. Finally, validation of these simulated field experiments with processes occurring in natural ecosystems will provide data necessary for informed risk assessment of ecological effects. Continued research (basic and applied) on a broad front will result in improvements to the general mesocosm design and postexperiment analyses.

As Cairns[81] succinctly stated:

> If environmental toxicology is to come of age, it must begin to ask more searching questions, develop broader hypotheses involving natural systems, and develop models that are validated in landscapes, not laboratories.

DISCLAIMER

The views expressed in this article are those of the author(s) and do not necessarily reflect a policy position of the U.S. Environmental Protection Agency.

REFERENCES

1. Voshell, J. R., Jr., Ed., Using Mesocosms for Assessing the Aquatic Ecological Risk of Pesticides: Theory and Practice, Entomological Society of America, Miscellaneous Publications No. 75 (1989), pp. 88.

2. Voshell, J. R., Jr., Introduction and Overview of Mesocosms, in *Experimental Ecosystem Applications to Ecotoxicology,* North American Benthological Society, Technical Information Workshop, Virginia Polytechnic Institute and State University, Blacksburg, VA (1990), pp. 1–3.

3. Cuffney, T. F., D. D. Hart, K. C. Wolbach, J. B. Wallace, G. J. Lugthart, and F. L. Smith-Cuffney, Assessment of Community and Ecosystem Level Effects in Lotic Environments: The Role of Mesocosm and Field Studies, in *Experimental Ecosystems: Applications to Ecotoxicology,* North American Benthological Society, Technical Information Workshop, Virginia Polytechnic Institute and State University, Blacksburg, VA (1990).

4. Graney, R. L., J. H. Kennedy, and J. H. Rodgers, Jr, Eds., *Environmental Toxicology and Chemistry,* Society of Environmental Toxicology and Chemistry, Special Publications Series, Lewis Publishers, Boca Raton, FL (1993).

5. Aquatic Effects Dialogue Group (AEDG), Ed., Improving Aquatic Risk Assessment Under FIFRA, in *Report of the Aquatic Effects Dialogue Group,* World Wildlife Fund (1992).

6. Giddings, J. M., Summary, in *Workshop on Aquatic Microcosms for Ecological Assessment of Pesticides,* The Society of Environmental Toxicology and Chemistry Foundation for Environmental Education, Pensacola, FL (1992), pp. 14–25.

7. European Workshop on Freshwater Field Tests (EWOFFT) Programme–Abstracts–Delegates, Potsdam, Germany (1992).

8. Lundgren, A., Model Ecosystems as a Tool in Freshwater and Marine Research, *Arch. Hydrobiol. (Suppl.)* 70:157–196 (1985).

9. Odum, E. P., The Mesocosm, *BioScience* 34:558–562 (1984).

10. Cairns, J., Jr., Putting the Eco in Ecotoxicology, *Reg. Toxicol. Pharm.* 8:226–238 (1988).

11. Giesy, J. P., Jr. and E. P. Odum., Microcosmology: Introductory Comments, in *Microcosms in Ecological Research,* J. P. Giesy, Jr., Ed., Dept. of Energy Symposium Series 52, Conf. 781101, National Technical Information Service, Springfield, VA (1980), pp. 1–13.

12. Kennedy, J. H., Z. B. Johnson, P. C. Johnson, and P. D. Wise, The Use of Microcosm and Mesocosms as Surrogate Aquatic Ecosystems in Ecotoxicological Research: A Review, in *Handbook of Ecotoxicology,* D. J. Hoffman, Ed., Lewis Publishers, Boca Raton, FL (1995).

13. Resh, V. H. and D. M. Rosenberg, Spatial-temporal Variability and the Study of a Aquatic Insects, *Can. Entomol.* 121:941–963 (1989).

14. Frost, T. M., D. L. DeAngelis, S. M. Bartell, D. J. Hall, and S. H. Hurlbert, Scale in the Design and Interpretation of Aquatic Community Research, in *Complex Interactions in Lake Communities,* S. R. Carpenter, Ed., New York, NY: Spring-Verlag (1988), pp. 229–258.

15. Giddings, J. M. and G. I. Eddlemon, The Efects of Microcosm Size and Substrate Type on Aquatic Microcosm Behavior and Arsenic Transport, *Arch. Environ. Contam. Toxicol.* 6:491–505 (1977).

16. Stephenson, G. L., P. Hamilton, N. K. Kaushik, J. B. Robinson, and K. R. Solomon, Spatial Distribution of Plankton in Enclosures of Three Sizes, *Can. J. Fish. Aquat. Sci.* 41:1048–1054 (1984).

17. Solomon, K. R., G. L. Stephenson, and N. K. Kaushik, Effects of Methoxychlor on Zooplankton in Freshwater Enclosures: Influence of Enclosure Size and Number of Applications, *Environ. Toxicol. Chem.* 8:659–669 (1989).

18. Dudzik, M., J. Harte, A. Jassby, E. Lapan, D. Levy, and J. Rees, Some Considerations in the Design of Aquatic Microcosms for Plankton Studies, *Int. J. Environ. Stud.* 13:125–130 (1979).

19. Johnson, P. C., Impacts of the Pyrethroid Insecticide Cyfluthrin on Aquatic Invertebrate Populations in Outdoor Experimental Tanks, Ph.D. Thesis, University of North Texas, Denton, TX (1992).

20. Johnson, P. C., J. H. Kennedy, R. G. Morris, F. E. Hambleton, and R. L. Graney, Fate and Effects of Cyfluthrin (Pyrethroid Insecticide) in Pond Mesocosms and Concrete Microcosms, in *Aquatic Mesocosm Studies in Ecological Risk Assessment,* R. L. Graney, J. H. Kennedy, and J. H. Rodgers, Jr., Eds., Special Publications Series, Society of Environmental Toxicology and Chemistry Lewis Publishers, Boca Raton, FL (1993).

21. Howick, G. L., F. deNoyelles, J. M. Giddings, and R. L. Graney, Earthen Ponds Versus Fiberglass Tanks as Venues for Environments: A Parallel Study with Sulprofos, in *Aquatic Mesocosm Studies in Ecological Risk Assessment,* R. L. Graney, J. H. Kennedy, and J. H. Rodgers, Jr., Eds., Special Publications Series, Society of Environmental Toxicology and Chemistry Lewis Publishers, Boca Raton, FL (1993).

22. Allan, J. D., Hypothesis Testing in Ecological Studies of Aquatic Insects, in *The Ecology of Aquatic Insects,* V. H. Resh and D. M. Rosenberg, Eds., New York, NY: Praeger (1984), pp. 484–507.

23. Crow, M. E. and F. B. Taub, Designing a Microcosm Bioassay to Detect Ecosystem Level Effects, *Int. J. Environ. Stud.* 13:141–147 (1979).

24. O'Neil, P. E., S. C. Harris, and M. F. Mettee, Experimental Stream Mesocosms as Applied in the Assessment of Produced Water Effluents Associated with the Development of Coalbed Methane, *Experimental Ecosystems: Application to Ecotoxicology,* North American Benthological Society, Technical Information Workshop, Virginia Polytechnic Institute and State University, Blacksburg, VA (1990).

25. Solomon, K. R., Lake Mesocosms: Techniques and Procedures for Assessing Pesticide Ecotoxicology, in *Experimental Ecosystems: Application to Ecotoxicology,* North American Benthological Society, Technical Information Workshop, Virginia Polytechnic Institute and State University, Blacksburg, VA (1990).

26. Crossland, N. O. and T. W. La Point, The Design of Mesocosm Experiments, *Environ. Toxicol. Chem.* 11:1–4 (1992).

27. Dickson, K. L. and J. Cairns, Jr., The Relationship of Freshwater Macroinvertebrate Communities Collected by Floating Artificial Substrates to the MacArthur-Wilson Equilibrium Model, *Amer. Midl. Nat.* 88:68–75 (1972).

28. Needham, P. R. and R. L. Usinger, Variability in the Macrofauna of a Single Riffle in Prosser Creek, California, as Indicated by the Surber Sampler, *Hilgardia* 24:383–409 (1956).

29. Chutter, F. M. and R. G. Noble, The Reliability of a Method of Sampling Stream Invertebrates, *Arch. Hydrobiol.* 62:95–103 (1966).

30. Pratt, J. R. and N. J. Bowers, Variability of Community Metrics: Detecting Changes in Structure and Function, *Environ. Toxicol. Chem.* 11:451–457 (1992).

31. Schindler, D. W., Detecting Ecosystem Responses to Anthropogenic Stress, *Can. J. Fish. Aquat. Sci.* 44:6–25 (1987).

32. Maciorowski, A. F., Populations and Communities: Linking Toxicology and Ecology in a New Synthesis, *Environ. Toxicol. Chem.* 7:677–678 (1988).

33. Crossland, N. O. and D. Bennett, Fate and Biological Effects of Methyl Parathion in Outdoor Ponds and Laboratory Aquaria, II, Effects, *Ecotoxicol. Environ. Saf.* 8:482–495 (1984a).

34. Crossland, N. O. and D. Bennett, Fate and Biological Effects of Methyl Parathion in Outdoor Ponds and Laboratory Aquaria, II, Effects, *Ecotoxicol. Environ. Saf.* 8:482–495 (1984b).

35. Crossland, N. O., D. Bennett, C. J. M. Wolff, and R. P. J. Swannell, Evaluation of Models to Assess the Fate of Chemicals in Aquatic Systems, *Pestic. Sci.* 17:297–304 (1986).

36. Heimbach, F., W. Pflueger, and H-T. Ratte, Use of Small Artificial Ponds for Assessment of Hazards to Aquatic Ecosystems, *Environ. Toxicol. Chem.* 11:27–34 (1992).

37. Kennedy, J. H., Z. B. Johnson, and P. C. Johnson, Sampling and Analysis Strategy for Biological Effects in *Freshwater Field Tests for Hazard Assessment of Chemicals,* I. R. Hill, F. Heimbach, P. Leeuwangh, and P. Matthiessen, Eds., Lewis Publishers, Boca Raton, FL (1995).

38. Wolff, C. J. M. and N. O. Crossland, Fate and Effects of 3, 4-Dichloroaniline in the Laboratory and in Outdoor Ponds: I, Fate, *Environ. Toxicol. Chem.* 4:481–487 (1985).

39. Howick, G. L., J. M. Giddings, F. deNoyelles, L. C. Ferrington, Jr., W. D. Kettle, and D. Baker, Rapid Establishment of Test Conditions and Trophic-level Interactions in 0.04-hectare Earthen Pond Mesocosms, *Environ. Toxicol. Chem.* 11:107–114 (1992).

40. Touart, L. W., Aquatic Mesocosm Testing to Support Pesticide Registration, Office of Pesticide Programs, U.S. EPA 540/09-88-035 (1987).

41. Arumugam, P. T. and M. C. Geddes, An Enclosure for Experimental Field Studies with Fish and Zooplankton Communities, *Hydrobiologia* 135:215–221 (1986).

42. Brooks, J. L. and S. I. Dodson, Predation, Body Size, and Composition of Plankton, *Science* 150:28–35 (1965).

43. Drenner, R. W., S. T. Threlkeld, and M. D. McCracken, Experimental Analysis of the Direct and Indirect Effects of an Omnivorous Filter-feeding Clupeid on Plankton Community Structure, *Can. J. Fish. Aquat. Sci.* 43:1935–1945 (1986).

44. Vinyard, G. L., R. W. Drenner, M. Gophen, U. Pollingher, D. L. Winkleman, and K. D. Hambright, An Experimental Study of the Plankton Community Impacts of Two Omnivorous Filter-feeding Cichlids, *Tilapia galilaea* and *Tilapia aurea*, *Can. J. Fish. Aquat. Sci.* 45:685–690 (1988).

45. Mazumder, A., D. J. McQueen, W. D. Taylor, and D. R. S. Lean, Effects of Fertilization and Planktivorous Fish (Yellow Perch) Predation on Size Distribution of Particulate Phosphorus and Assimilated Phosphate: Large Enclosure Experiments, *Limnol. Oceanogr.* 33:421–430 (1988).

46. Mazumder, A., W. D. Taylor, D. J. McQueen, and D. R. S. Lean, Effects of Fertilization and Planktivorous Fish on Epilimnetic Phosphorus and Phosphorus Sedimentation in Large Enclosure, *Can. J. Fish. Aquat. Sci.* 46:1735–1742 (1989).

47. Threlkeld, S. T., Planktivory and Planktivore Biomass Effects on Zooplankton, Phytoplankton, and the Trophic Cascade, *Limnol. Oceanogr.* 33:1326–1375 (1988).

48. Brabrand, Å., B. Faafeng, and J. P. M. Nilssen, Pelagic Predators and Interfering Algae: Stabilizing Factors in Temperate Eutrophic Lakes, *Arch. Hydrobiol.* 110:533–552 (1987).

49. Langeland, A., J. I. Koksvik, Y. Olsen, and H. Reinertsen, Limnocorral Experiments in a Eutrophic Lake—Effects of Fish on the Planktonic and Chemical Conditions, *Pol. Arch. Hydrobiol.* 34:51–65 (1987).

50. Hill, I. R., S. T. Hadfield, J. H. Kennedy, and P. Ekoniak, Assessment of the Impact of PP321 on Aquatic Ecosystems using Tenth-acre Experimental Ponds, in Brighton Crop Protection Conference — Pests and Diseases (1988), pp. 309–318.

51. Hill, I. R., J. K. Sadler, J. H. Kennedy, and P. Ekoniak, Lambda-cyhalothium: A Mesocosm Study of its Effects on Aquatic Organisms, in *Aquatic Mesocosm Studies in Ecological Risk Assessment,* R. L. Graney, J. H. Kennedy, and R. H. Rodgers, Jr., Eds., Special Publications Series, Society of Environmental Toxicology and Chemistry Lewis Publishers, Boca Raton, FL (1993).

52. Morris, R. G., J. H. Kennedy, and P. C. Johnson, Comparison of the Effects of the Pyrethroid Insecticide Cyfluthrin on Bluegill Sunfish, in *Aquatic Mesocosm Studies in Ecological Risk Assessment,* R. L. Graney, J. H. Kennedy, and R. H. Rodgers, Jr., Eds., Special Publications Series, Boca Raton, FL: Lewis Publishers (1993).

53. Mayashich, J. M., J. H. Kennedy, and J. S. O'Grodnick, Evaluation of the Effects of Tralomethrin on Aquatic Ecosystems, in *Aquatic Mesocosm Studies in Ecological Risk Assessment,* R. L. Graney, J. H. Kennedy, and R. H. Rodgers, Eds., Special Publications Series, Boca Raton, FL: Lewis Publishers (1993).

54. Carpenter, S. R., J. F. Kitchell, and J. R. Hodgson, Cascading Trophic Interactions and Lake Productivity, *BioScience* 35:634–639 (1985).

55. Threlkeld, S. T. Experimental Evaluation of Trophic-cascade and Nutrient-mediated Effects of Planktivorous Fish on Plankton Community Structure, in *Predation: Direct and Indirect Impacts on Aquatic Communities,* W. C. Kerfoot and A. Sih, Eds., University Press of New England (1987), pp. 161–173.

56. McQueen, D. J. and J. R. Post, Cascading Trophic Interactions: Uncoupling at the Zooplankton-phytoplankton Link, *Hydrobiologia* 159:277–296 (1988).

57. Deutsch, W. G., E. C. Webber, D. R. Bayne, and C. W. Reed, Effects of Largemouth Bass Stocking Rate on Fish Populations in Aquatic Mesocosms Used for Pesticide Research, *Environ. Toxicol. Chem.* 11:5–10 (1992).

58. Stunkard, C. and T. Springer, Statistical Anslysis and Experimental Design, in *Imporving Aquatic Risk Assessment Under FIFRA Report of the Aquatic Effects Dialogue Group,* A.D.D.G., Eds., World Wildlife Fund (1992), pp. 65–75.

59. Drenner, R. W., K. D. Hoagland, J. D. Smith, W. J. Barcellona, P. C. Johnson, M. A. Palmieri, and J. F. Hobson, Effects of Sediment-bound Bifenthrin on Gizzard Shad and Plankton in Experimental Tank Mesocosms, *Environ. Toxicol. Chem.* 12:1297–1306 (1993).

60. Brazner, J. C., L. J. Heinis, and D. A. Jensen, A Littoral Enclosure for Replicated Field Experiments, *Environ. Toxicol. Chem.* 8:1209–1216 (1989).

61. Yasuno, M., T. Hanazato, T. Iwakuna, K. Takamura, R. Ueno, and N. Takamura, Effects of Permethrin on Phytoplankton and Zooplankton in an Enclosure Ecosystem in a Pond, *Hydrobiologia* 159:247–258 (1988).

62. Liber, K., N. K. Kaushik, K. R. Solomon, and J. H. Carey, Experimental Designs for Aquatic Mesocosm Studies: A Comparison of the "ANOVA" and "Regression" Design for Assessing the Impact of Tetrachlorophenol on Zooplankton Populations in Limnocorrals, *Environ. Toxicol. Chem.* 11:61–77 (1992).

63. Ludwig, J. A. and J. F. Reynolds, *Statistical Ecology,* New York, NY: John Wiley and Sons (1988).

64. Green, R. H., *Sampling Design and Statistical Methods or Environmental Biologists,* New York, NY: John Wiley and Sons (1979).

65. Boyle, T. B., G. M. Smillie, J. C. Anderson, and D. R. Beeson, A Sensitivity Analysis of Nine Diversity and Seven Similarity Indices, *Res. J. Water Pollut. Control Fed.* 62(6):749–762 (1990).

66. Nemec, A. F. L. and R. O. Brinkhurst, Using the Bootstrap to Assess Statistical Significance in the cluster Analysis of Species Abundance Data, *Can. J. Fish. Aquat. Sci.,* 45:965–970 (1988).

67. Bray, R. and J. T. Curtis, An Ordination of the Upland Forest Communities of Southern Wisconsin, *Ecol. Monogr.* 27:325–349 (1957).
68. Diaconis, P. and B. Efron, Computer-intensive Methods in Statistics, *Sci. Am.* 248:116–130 (1983).
69. Efron, B., Bootstrap Methods: Another Look at the Jacknife, *Ann. Statistics* 7:1–26 (1979).
70. Felsenstein, J., Confidence Limits on Phylogenies: An Approach Using the Bootstrap, *Evolution* 39:783–791 (1985).
71. Pontasch, K. W., B. R. Niederlehner, and J. Cairns, Jr., Comparisons of Single Species Microcosm and Field Responses to a Complex Effluent, *Environ. Tox. Chem.* 8:521–522 (1989).
72. Karr, J. R., Assessment of Biotic Integrity Using Fish Communities, *Fisheries* 6:21–27 (1981).
73. Karr, J. R., K. D. Fausch, P. L. Angermeier, P. R. Yant, I. J. Schlosser, Assessing Biological Integrity in Running Waters: A Method and its Rationale, Special Publication 5, Illinois Natural History Survey, Champaign, IL, 28 (1986).
74. Plafkin, J. L., M. T. Barbour, K. D. Porter, S. K. Gross, and R. M. Hughes, Rapid Bioassessment Protocols for Use in Streams and Rivers: Benthic Macroinvertebrates and Fish, U.S. E.P.A. 444/4-89-001 (1989).
75. Cairns, J., Jr., What is Meant by Validation of Predictions Based on Laboratory Toxicity Tests? *Hydrobiologia* 137:271–278 (1986).
76. Cairns, J., Jr., What Constitutes Field Validation of Predictions Based On Laboratory Evidence? in *Aquatic Toxicology and Hazard Assessment: Tenth Volume,* ASTM STP 971, W. J. Adams, G. A. Chapman, and W. G. Landis, Eds., American Society for Testing and Materials, Philadelphia, PA (1988), pp. 361–368.
77. Cairns, J., Jr. and E. P. Smith, Developing a Statistical Support System for Environmental Hazard Evaluation, *Hydrobiologia* 184:143–151 (1989).
78. Mount, D. I., Scientific Problems in Using Multispecies Toxicity Tests for Regulatory Purposes, in *Multispecies Toxicity Testing,* J. Cairns, Jr., Ed., New York: Pergamon Press (1985), pp. 13–18.
79. La Point, T. W. and J. A. Perry, Use of Experimental Ecosystems in Regulatory Decision Making, *Environ. Manag.* 13:539–544 (1989).
80. Golley, F. B., What Do Ecologists Expect from Industry? in *Multispecies Toxicity Testing,* J. Cairns, Jr., Ed., New York, NY: Pergamon Press (1985), pp. 27–35.
81. Cairns, J., Jr., Paradigms Flossed: The Coming of Age of Environmental Toxicology, *Environ. Toxicol. Chem.* 11:285–287 (1992).

CHAPTER 10

The Use of Stream Microcosms in Multispecies Testing

Kurt W. Pontasch

> One thing to my mind was certain, that to make any laboratory experiments that would compare in any measure to what goes on in a river, it was absolutely necessary that the water experimented upon should be *running*, and not merely exposed to light and air in a bottle.

> C. M. Tidy, 1880

INTRODUCTION

HISTORICAL OVERVIEW

The use of artificial streams as models for understanding ecological processes and the fate and effects of contaminants began in England more than a century ago with Tidy's[1] studies into the "self-purification" of sewage. Apparently, Tidy's work with artificial streams failed to interest his contemporaries because further work was not reported until the middle part of this century when artificial streams were once again employed to study the fate of sewage inputs[2–4] and the safety and efficacy of several insecticides for controlling black flies.[5] Interest in artificial stream research increased in the 1960s, and much of that basic research has been previously reviewed.[6–8] Artificial streams were also used during the 1960s to study applied questions such as the effects of temperature,[9–10] surfactants,[11] and insecticides[12] on stream organisms. Also, at this time, Rose and McIntire[13] were likely the first to utilize model streams to study accumulation of a toxicant (dieldrin) by periphytic communities. Based on these early studies, the U.S. Environmental Protection Agency, in

1973, tentatively recommended a periphyton bioassay utilizing artificial streams.[14] Most of the research outlined above was conducted by a few research groups with rather dissimilar interests and objectives, and, in some cases, the test organisms represented only one or two species rather than communities of organisms that inhabit natural streams. It was not until the late 1970s that the use of artificial streams for multispecies testing began to spread.

Maki and coworkers conducted one of the most extensive studies in the late 1970s with the lampricide, TFM (3-trifluoromethyl-4-nitrophenol). They used artificial streams to determine TFM's toxicity to macroinvertebrates[15] and effects on primary production and community respiration.[16] Further studies indicated that the model streams reflected responses in a natural stream exposed to TFM.[17] Also in the late 1970s, Burks and Wilhm,[18] in their studies of petroleum refinery wastewaters, were probably the first to utilize colonized artificial substrates as a source of test organisms to stock model streams. Experiments conducted in the "channels microcosm" at the Savannah River Plant determined the effects of mercury[19] and cadmium[20] on stream benthos. Other workers in the 1970s used artificial streams to investigate the effects of coal leachates,[21] and kraft mill,[22] and municipal treatment plant[23] effluents on various stream communities. The results of a number of other artificial stream studies conducted at this time were published[24] following a symposium on "Microcosms in Ecological Research".

From 1980 to the present, the amount of published research employing artificial streams to address the fate and/or effects of contaminants has increased considerably (Table 1).[25–70] That trend has probably resulted both from an increase in the number of ecotoxicologists and a heightened awareness of the inability of standard, single-species tests to predict contaminant effects on ecosystem structure and function.[71–73] At the present time, interest in artificial stream research, and community level testing in general, is still expanding as evidenced by the number of special symposia and workshops in recent years. Pond mesocosm studies have been used in support of pesticide registrations under the Federal Insecticide, Fungicide and Rodenticide Act (FIFRA), and those methodologies are being standardized.[74,75] However, the role of stream microcosm and mesocosm studies is not as well defined.

THE ROLE OF STREAM MICROCOSMS AND MESOCOSMS IN ECOTOXICOLOGY

What constitutes a stream microcosm vs. a stream mesocosm has generally not been agreed upon. The distinction made by Odum[76] that microcosms are laboratory based, and mesocosms are outdoor facilities will be employed throughout this paper. However, it should be noted that using this distinction results in outdoor "mesocosms" that are, in some cases, much smaller than many indoor "microcosms." There are advantages and disadvantages associ-

Table 1 Number and Types of Studies Employing Artificial Streams to Determine the Fate and Effects of Environmental Contaminants

Years	1880–1959	1960–1969	1970–1979	1980–1992	Refs.
Laboratory Microcosms					
Fate			1	3	13, 36, 38, 40
Effects					
Invertebrates		2	2	6	10, 11, 15, 22, 25–27, 30–32
Periphyton		1	1	4	9, 21, 33–35, 37
Invertebrates and periphyton			1	3	16, 28, 29, 39
Subtotal		3	5	16	
Outdoor Mesocosms					
Fate	4		1	3	1–4, 20, 65–67
Effects					
Invertebrates	1	1	1	13	5, 12, 18, 41–46, 48–54
Periphyton			1	13	23, 55–64, 68–70
Invertebrates and periphyton			2	1	19, 20, 47
Subtotal	5	1	5	30	
TOTAL	5	4	10	46	

ated with both stream microcosms and mesocosms, but, before discussing those, the role of artificial stream research in ecotoxicology will be addressed.

Most municipal and industrial effluents are discharged into streams and rivers, and lotic ecosystems of all sizes are exposed to agricultural contaminants. However, data currently used in assessing chemical hazards in lotic ecosystems are derived from test systems that model ponds and lakes. The passage at the beginning of this chapter suggests that, over 100 years ago, Tidy[1] understood that laboratory tests conducted in a beaker could not model the ecological attributes of flowing waters and were, therefore, inadequate predictors of the fate and effects of contaminants that enter streams and rivers. The single species tests currently employed in aquatic hazard assessments have proven to be valuable tools for determining relative toxicity, potential for bioaccumulation, behavioral effects, reproductive success, and other endpoints. Such tests are probably sufficient for routine screening of chemicals that are not expected to enter the environment under normal circumstances. However, single species tests may be overprotective when large application factors are employed, and accurate predictions of the extent and type of deleterious effects are not possible. Therefore, in evaluating pesticides and contaminants that will be released in municipal and industrial effluents, the use of stream microcosm or mesocosm tests should be considered.

One of the benefits of stream microcosms and mesocosms is that they allow direct assessments of effects on communities indigenous to potentially impacted streams. Consequently, validation of predictions is less difficult because prediction and validation are conducted at comparable levels of biological organization. For example, when establishing permits for a new effluent or changes in an existing effluent, multispecies tests with communities indigenous to the receiving stream allow direct predictions of effects and subsequent field validation of those predictions.[31] Similarly, multispecies tests with indigenous organisms would help avoid the costs associated with construction and operation of new waste treatment facilities that provide no biological benefit. During routine evaluations of existing effluents, upstream-downstream biosurveys in conjunction with single species tests to detect effluent variability are probably more appropriate than multispecies tests.

The effects of nonpoint source agricultural pollutants usually cannot be detected by biosurveys because suitable reference sites are not present. In addition, as previously noted, registration of new pesticides is based on the results of tests that model lentic ecosystems. Therefore, in most cases, little or no data concerning the effects of pesticides on stream communities are available for pesticides currently in use. Small, partially recirculating laboratory microcosms could easily be employed to test the effects of pesticides on stream communities. Such tests should cost no more than some of the more elaborate single species tests currently used and would be less expensive than pond mesocosm studies. The microcosms should model the physical, chemical, and biological characteristics of streams in the ecoregion where the pesticide will be used. If streams within a given ecoregion vary in terms of morphometric characteristics, water chemistry, and species assemblages, these attributes could be controlled in a series of tests within the same laboratory.

Because natural streams are open systems, larger, once-through microcosms[36,38,40] and mesocosms[20,65] may be more appropriate than partially recirculating microcosms for predicting the ultimate fate of pesticides and other chemicals. Such streams should be open channels of sufficient size to model riffles, runs, and pools and have retention times that allow for environmental partitioning and chemical degradation. Unfortunately, these requirements often result in insufficient or no replication of experimental units because of the increased costs associated with establishing and maintaining large, complex artificial streams.

In addition to increased cost and inadequate replication, large stream mesocosms are inherently more variable than smaller stream microcosms. Large mesocosms may model natural streams better than microcosms, but there is a significant decrease in control of many experimental variables. For example, light intensity, temperature, and water chemistry will change with the weather in exposed outdoor mesocosms, and periods of intense precipitation can reduce test material concentrations. It is also more difficult to quantify changes in test community structure and function because the distribution of test organisms in large mesocosms is often as patchy as that of heterogeneous natural environments. In addition, quantification of aquatic insect emergence (discussed below) is much more difficult. Therefore, relative to smaller microcosms, the number of replicate treatments and samples per replicate required for proper statistical analysis of treatment responses are substantially greater. Conversely, treatment responses in smaller microcosms are more easily interpreted because cause and effect are not as confounded. Another disadvantage of large mesocosms, and some microcosms, is that they are site specific and, therefore, lack flexibility in terms of communities tested and conditions of exposure. Because of the above considerations, laboratory microcosms appear to offer the best prospect for incorporation into the regulatory framework, and therefore, the rest of this chapter will focus primarily on stream microcosms.

Before stream microcosms can begin to fill some of the roles outlined above and be incorporated into the regulatory framework, a limited set of community and ecosystem level endpoints should be selected and tested for reproducibility and predictive utility. In addition, the various microcosm designs and methods for obtaining test organisms need to be somewhat standardized. However, experimental conditions such as temperature, water chemistry, illumination, and current velocity should remain flexible and reflect conditions in the source ecosystem. Some of the methods and endpoints developed for stream microcosms are discussed below within the context of their prospective role in ecotoxicology.

STREAM MICROCOSMS

DESIGNS

Stream microcosm designs vary in terms of size, shape, construction material, substrate type, source of current, water flow, etc. and have been previously reviewed.[77] The design employed depends on the research objectives and type of community being investigated. For example, substrates used in tests with stream macroinvertebrates are generally a mixture of various sizes of river rock intended to model the natural substrate (Table 2). Conversely, tests with periphytic communities often successfully employ artificial substrates that bear no resemblance to the natural substrate (Table 3). In all cases, the construction material should be nontoxic and minimize adsorption of test material to stream walls. A number of laboratories are currently using oval, fiberglass streams, first developed by Pontasch and Cairns,[30] that are inexpensive, lightweight, portable, and easy to clean. Microcosms used to study contaminant fate and/or effects on microbial communities have generally been longer, once-through channels with water and current supplied by diverting a portion of a natural stream into the laboratory.[33,34,36–40] Such microcosms have also been used to study contaminant effects on macroinvertebrate communities,[15,16,29] but

Table 2   Stream Microcosm Designs Used in Multispecies Tests with Macroinvertebrate Communities[a]

Water Flow	Shape	Size (m) L–W–D[b]	Number	Current Source	Construction Material	Substrate Type	Ref.
Partially recirculating	Oval	0.9–0.1–0.1	10	Paddle wheels	Iron coated with epoxy resin	Flat stones	11
Partially recirculating	Oval	0.7–0.3–0.1	12	Pump/paddle wheels	Stainless steel	"Natural"	10
Partially recirculating	Rectangle	3.3–0.7–0.3	6	Paddle wheels	Wood	Gravel and rubble	22
Once-through	Trough	8.0–0.6–0.3/0.1	6	Gravity	Concrete	Gravel and rubble	15,16
Partially recirculating	Oval	0.8–0.5–0.1	15	Paddle wheels	Wood/nontoxic paint	Containers of pebble/cobble	25
Partially recirculating	Oval	0.8–0.5–0.1	12	Paddle wheels	Wood/nontoxic paint	Containers of pebble/cobble	26
Partially recirculating	Rectangle	12.2–0.3–0.3	10	Paddle wheels	NA[c]	7.5–13.0 cm rocks	28
Once-through	Trough	4.0–0.6–0.3	6	Gravity	Concrete	"Natural"	29,36
Partially recirculating	Oval	1.7–0.2–0.1	16	Paddle wheels	Fiberglass	Containers of pebble/cobble	32

[a] References 16, 28, and 29 also tested microbial communities
[b] Length, width, depth; dimensions are rounded to the nearest 0.1 m
[c] NA = Not available

Table 3   Stream Microcosm Designs Used in Multispecies Tests with Microbial Communities

Water Flow	Shape	Size (m) L–W–D[a]	Number	Current Source	Construction Material	Substrate Type	Ref.
Partially recirculating	Rectangle	3.0–0.3–0.2	12	Paddle wheels	Wood/nontoxic paint	Smooth rocks	13
Once-through	Trough	11.0–0.4-NA[b]	4	Gravity	NA	Clay tiles	38
Once-through	Trough	20.0–0.3–0.3	1	Gravity	NA	"Sediment"	40
Once-through	Trough	19.5-NA-0.5	4	Gravity	Teflon-lined plexiglass	Teflon coated plastic	33
Once-through	Trough	19.5-NA-0.5	2	Gravity	Teflon-lined plexiglass	Teflon coated plastic	34
Partially recirculating	Tube	0.1-NA-NA	6	Pump	Glass	Glass	35
Recirculating	Trough	12.0–1.5–0.1/0.3	2	Pump	Concrete/epoxy paint	Natural/glass slides	37

[a] Length, width, depth; dimensions are rounded to the nearest 0.1 m
[b] NA = Not available

smaller, partially recirculating microcosms are used more often in effects studies with macroinvertebrates (Table 2).

Stream microcosms designed to determine effects on macroinvertebrate communities are usually similar to the laboratory stream system developed in 1963 by Surber and Thatcher[11] in that they are partially recirculating, oval streams with current supplied by paddle wheels. However, there is considerable variability in size, construction material, and substrate type (Table 2). Although larger microcosms can provide a more complex environment by including riffle and pool areas and/or a variety of substrate sizes, reproducibility of test results may decrease as complexity increases.

Macroinvertebrate species richness is much higher in riffles than pools of natural streams, and, in most streams and rivers, aquatic insects constitute the majority of both species and individuals. Riffles also contain a higher percentage of Ephemeroptera, Plecoptera, and Trichoptera (Insecta) species that are generally more sensitive to contaminants than other macroinvertebrates. Conversely, natural pools contain macroinvertebrate communities that are generally more tolerant of contaminants. Therefore, stream microcosm designs should, at a minimum, provide an environment suitable for maintaining and testing riffle insect communities.

Aquatic insects, perhaps due to lower resistance during molting, tend to exhibit increased sensitivity with longer exposures.[15,78,79] In addition, several studies have suggested that earlier instars are more sensitive than later instars to pesticides[15,80,81] and heavy metals.[78,82,83] Consequently, to conduct ecologically meaningful bioassays with riffle insects, a relatively long exposure period is necessary to insure that most species undergo one or more molts and that some species have complete life cycles during testing. For most riffle insect communities, a 30-day exposure period should satisfy both of these requirements.

Pontasch and Cairns[30] conducted a study to determine which of the following stream microcosm designs would be most logistically simple yet effective in maintaining riffle insect communities during a 30-day bioassay: (1) static and no current (S-NC); (2) flow-through and no current (FT-NC); (3) static with current (S-C); and (4) flow-through with current (FT-C). Flow-through and current, when provided, where 12 ml min[-1] and 30 cm sec[-1], respectively. Over the 30-day test, the FT-C and S-C designs maintained most taxa at or above initial densities, and even in the FT-NC and S-NC streams, densities of some taxa were not significantly different from initial densities. This study indicated that microcosms of riffle insect communities can be maintained for at least 30 days with moderate current and minimal flow-through.

Two other important considerations when designing stream microcosms for use in determining effects on macroinvertebrate communities are the immigration and emigration of organisms during testing. Immigration and emigration occur continuously in natural riffles, and it can be argued that stream microcosms should be designed to allow immigration via the inflow (assuming the dilution water is from a nearby stream) and emigration via the outflow. Such a system probably allows structural and functional treatment responses in the microcosms better to reflect those that would occur in a natural community. However, if immigration is not controlled, the exposure period is unknown, and test results may be incorrectly interpreted. Similarly, macroinvertebrates that leave the microcosm through the standpipe or end of the trough because of contaminant induced drift will be incorrectly counted as mortalities. Therefore, microcosms used to predict no observable effects concentrations (NOECs) for macroinvertebrate communities should either prevent or quantify immigration and emigration of organisms during testing.

Another factor that has confounded data interpretation since Surber and Thatcher's[11] pioneering work is the emergence of adult insects during toxicity tests. Recent studies have shown that a large proportion of initial, riffle insect densities may emerge during testing.[30–32] Toxicity tests with aquatic insects require monitoring of adult emergence so that losses due to emergence are not counted as mortalities and so that any toxicant-induced inhibition of adult emergence[cf.78] can be detected. Although most stream microcosms have not been designed to collect emergent adult insects, emergence traps should be employed in all studies that include aquatic insects as test organisms.

EXPERIMENTAL CONDITIONS

In contrast to single species tests, conditions of exposure in stream microcosms have not been standardized. The sensitivity of stream communities to changes in abiotic conditions is, in most cases, poorly understood, so microcosm temperature, current velocity, light intensity and quality, photoperiod, and water chemistry should reflect conditions in the stream from which test organisms were obtained. Further research is needed to determine acceptable ranges for these abiotic variables and their effects on different stream communities. Variables that could be somewhat standardized are the duration of exposure and rate of water replacement.

The stream used as a source of test organisms would provide the best duplication of water chemistry. However, if the source stream is some distance from the laboratory, the logistics and cost of transporting large quantities of dilution water prevent this approach. Dechlorinated tap water has been used successfully in studies with macroinvertebrates (Table 4) and deionized water has been used in some studies with microbial communities (Table 5). In one of the few studies that compared microcosm water chemistry to that of the source ecosystem, Pontasch and Cairns[30] found that most macroinvertebrate taxa were maintained at or above initial densities over a 30-day test period even though conductivity, alkalinity, and hardness in the microcosms were less than half that of the source ecosystem. This finding suggests that microcosm water chemistry can be used as a treatment variable during toxicity testing with macroinvertebrate communities. In addition, stream microcosm water chemistry

Table 4   Experimental Conditions in Stream Microcosms During Tests with Macroinvertebrate Communities[a]

Water Source	Conductivity (μmhos cm⁻¹)	pH	Alkalinity (mg L⁻¹ CaCO₃)	Hardness (mg L⁻¹ CaCO₃)	DO (mg L⁻¹)	Temp. (°C)	Current (cm sec⁻¹)	Light Source	Ref.
Spring		8.1–8.3	209–282	280–342	5.7–7.5	18.5–26.0		Fluorescent 12:12 L:D	11
DT[b]		7.2–7.5	40	35–40	>90% saturation	Variable		Translucent roof	10
Stream unfiltered							24 riffles 10 pools		22
Stream unfiltered		7.8	179	211		5.0–13.0		Fluorescent NP[c]	16
DT	115–173	7.9–8.4	41–66	64–77		8.3–20.8	35	Fluorescent 12:12 L:D	25
River unfiltered		8.6	64	88		14.0–15.0	38	Fluorescent 12:12 L:D	26
Well		6.9–7.2	2–3	25–61	9.9–11.3	10.2–14.2	30	Metal halide lamps NP	28
Stream unfiltered		7.4–7.8	164–202	182–217	9.0–11.2	15.5–0.5		Fluorescent NP	29
DT	123	8.2		45	8.6	18.0	25	Grow lights NP	32

[a] References 16, 28, and 29 also tested microbial communities; missing values are not available

[b] DT = Dechlorinated tap water

[c] NP = Natural photoperiod

Table 5 Experimental Conditions in Stream Microcosms During Tests with Microbial Communities[a]

Water Source	Conductivity (μmhos cm⁻¹)	pH	Alkalinity (mg L⁻¹ CaCO₃)	Hardness (mg L⁻¹ CaCO₃)	DO (mg L⁻¹)	Temp. (°C)	Current (cm sec⁻¹)	Light Source	Ref.
Stream filtered	175–214	7.3–7.7		72–98			28 or 0	Translucent roof	13
River unfiltered		7.4				12.0–21.0	56	Metal halide lamps NP[b]	38
River/well unfiltered		7.5–8.0	74	125	8.0	20.0–25.0	0.025	Fluorescent 8:16 L:D	40
Deionized						15.0		Fluorescent 12:12 L:D	33
Well and deionized						20.0		Fluorescent 12:12 L:D	34
Stream unfiltered								Fluorescent continuous	35
Deionized		7.3–7.4	13–17	11–15		3.0–16.0		Fluorescent NP	37

a Missing values are not available

b NP = natural photoperiod

can be altered as needed by relatively simple means and can, therefore, reflect conditions in any number of ecosystems used as a source of test organisms. The ability to model a variety of natural streams with the same system of stream microcosms provides flexibility not found with most mesocosms.

Temperature and photoperiod often influence the life histories of stream organisms. For example, these variables can affect the growth and emergence of aquatic insects[84] and have been shown to affect microbial community dynamics in laboratory streams.[6] Stream microcosms utilizing diverted stream water can reflect the seasonal and diel temperature fluctuations present in natural streams and permit a more natural exposure of test organisms to the variable of interest. However, microcosms maintained near the mean daily temperature of the source stream can support test communities for extended periods of exposure.[30-32] The photoperiod in some studies has corresponded to the natural photoperiod during testing; while other studies have utilized standardized photoperiods (Tables 4 and 5). Although it is tempting to attempt standardization of these variables, temperature and photoperiod in the stream microcosms should reflect natural conditions at the time of testing as closely as possible.

In most stream microcosm studies, light quality and intensity have not been given the attention they deserve. Although most laboratories have utilized fluorescent bulbs for lighting, lights designed to simulate the natural spectrum of sunlight may provide a better environment for primary production. In addition, stream microcosms generally model low-order streams that have some shading, but most microcosm designs lack shading, and the light source is directly above the microcosm. A light intensity of 50% of normal regional solar insolation has been recommended for stream microcosms,[85] but that level may be significantly higher than mean daylight intensities at the surface of shaded natural streams. Light quality and intensity are two additional variables that should reflect conditions in the source ecosystem.

The presence of a unidirectional current is one of the main factors distinguishing lotic from lentic ecosystems and results in a variety of microhabitats that control the distribution and abundance of both macroinvertebrate and microbial communities. However, most studies have either not reported microcosm current velocities or have not compared microcosm velocities to those of the source ecosystem. A microcosm current velocity of 30 cm sec^{-1} was sufficient for maintaining most macroinvertebrates taken from a natural riffle with current velocities that ranged from 50.3 to 98.5 cm sec^{-1}.[30] Reported microcosm current velocities during studies with macroinvertebrate communities (Table 4) are probably adequate for tests with most riffle macroinvertebrates. Current velocities have not been reported for many microcosm tests with microbial communities, and, in some cases, velocity has been kept artificially low to prevent sloughing of communities during testing (Table 5). However, microbial community structure is, in part, determined by current velocity,[86-88] and unreasonably low current velocities may result in microcosm microbial communities that do not resemble natural communities.

Nutrient availability is another factor that is often not considered during applied testing with macroinvertebrate and microbial communities. Functional feeding groups among stream macroinvertebrates include predators, shredders, collector–gatherers, and collector–filterers. Microcosms that are colonized by diverting a natural stream into the laboratory probably contain sufficient amounts of periphyton and coarse and fine particulate organic matter to provide an adequate macroinvertebrate food source. Microcosms that are stocked with macroinvertebrates previously colonized on artificial substrates (see below) should be seeded with a periphytic food source 4 to 6 weeks prior to introduction of macroinvertebrates. However, levels of filterable organic material may still be too low to maintain large populations of collector–filterers.[e.g.,30] Additions of coarse and fine particulate organic matter may be necessary to maintain shredder and collector–filterer populations, respectively. However, excess organic matter could reduce test material bioavailability. Additional research is necessary to determine the types and quantities of nutritional amendments that may be required during longer tests with macroinvertebrate communities.

In contrast to macroinvertebrates, the interactive effects of nutrient availability and herbivory on microbial community structure and function in stream microcosms are better understood.[e.g.,89,90] Overall, these studies suggest that nutrient (P and N) additions to stream microcosms are probably not necessary, even in partially recirculating streams but that the presence or absence of grazers may significantly affect results of tests with microbial communities. Applied tests with microbial communities that do not consider grazer effects may be poor predictors of fate and effects in natural streams.

TYPES AND SOURCES OF TEST COMMUNITIES

Multispecies tests with stream microcosms generally use microbial and macroinvertebrate communities. Although fish have been included in microcosm studies specifically designed to predict effects on fish production[22] and bioaccumulation through a simplified food chain,[27] most laboratory stream tests with fish have not included natural microbial and macroinvertebrate communities and are, therefore, not true microcosm tests. In microcosm studies where there is no immigration of macroinvertebrates, the presence of fish in sufficient numbers to statistically determine treatment responses may decimate macroinvertebrate populations and prevent assessments of effects on those organisms. Therefore, in most cases, fish should not be included in stream microcosms even though the absence of these top predators decreases environmental realism.

Microbial test communities are usually allowed to colonize stream microcosms via the replacement water supply that is diverted from a nearby stream or river. Other sources include natural stream[29,36,37,39] or pond[33] sediments and plankton tows from ponds[33] that are added to the microcosms. Colonized artificial substrates have also been used to seed microcosms with a food source prior to tests with macroinvertebrates.[30-32] The reported times allowed for microbial community development and equilibration prior to testing range from seven days to six months and depend on the source of organisms and the study objectives. Preliminary tests should quantify microbial community variability among replicate microcosms prior to testing. Although some tests with microbial communities have not included macroinvertebrate communities, as noted above, the absence of grazers may significantly alter test results. Therefore, it is suggested that grazers be included in stream microcosms even if treatment effects on that trophic group are not quantified.

Macroinvertebrate test communities in stream microcosms are usually established by: (1) allowing colonization through the replacement water supply; (2) using artificial substrates previously colonized in a natural stream; or (3) collecting organisms from a natural stream with a D net or similar sampling device. The first method requires a nearby stream with a diverse assemblage of macroinvertebrates, a relatively long colonization period, and, because of differences in drift behavior among species and life stages of a given species, may result in microcosm communities that do not reflect natural community structure and function. In addition, the microcosms must be large enough to allow preliminary sampling to quantify variability among replicate microcosms, and, as discussed above, immigration of test organisms during testing should be controlled or quantified.

The second method, colonization of artificial substrates placed in the natural ecosystem being modeled, requires a 3- to 4-week colonization period to reach maximum species richness and density equilibrium for most taxa. Variability in macroinvertebrate community structure in rock-filled artificial substrates is low,[91] and estimates of variability among replicate microcosms can be made by sampling colonized artificial substrates that will not be used in the stream microcosms. As with method (1) above, selective colonization of artificial substrates by some species may result in microcosm communities that are not representative of natural communities.[30,92]

Studies that employed method (3), collection of organisms directly from the natural substrate, only tested a portion of the macroinvertebrate community, and the organisms collected were all late instar insect larvae and nymphs. Although larger organisms are easier to collect and transport, as noted above, earlier instars are apparently more sensitive than later instars to some contaminants. Macroinvertebrate, microcosm communities that reflect the natural community in terms of species richness and the proportional contribution of the various species and life stages within a species to total numbers could be established by transporting intact quantitative samples (e.g., Hess or Surber samples) to the microcosms. However, the spatial heterogeneity of natural macroinvertebrate communities would probably cause increased variability among replicate microcosms relative to methods (1) and (2). In addition, although this method eliminates the colonization periods necessary with methods (1) and (2), the handling of organisms during sampling may result in mortality or reduced fitness prior to test initiation.

ENDPOINTS QUANTIFIED

Fate

Relatively few laboratories have used stream microcosms to study the environmental fate of chemical stressors. Most of the studies conducted measured accumulation in various biological components of stream microcosms such as microbial communities,[13,29,39] macrophytes,[29,39] macroinvertebrates,[27,29,39] and fish.[29,39] Other studies investigated the ability of microbial communities to adapt to and degrade the test chemical.[38,40] In addition, larger, once-through stream microcosms have been used to determine partitioning to sediments.[29,39] All of these studies appear valid and produced useful results. As noted above, larger, once-through microcosms and mesocosms are best suited for determining the ultimate fate of test chemicals. However, bioaccumulation studies have been successfully conducted in small, recirculating streams containing gnotobiotic assemblages of organisms[27] and small, partially recirculating streams containing natural assemblages of organisms.[13] The experimental design selected depends on the objectives of the research and type of chemical tested.

Effects

The majority of stream microcosm studies have measured toxicant effects on microbial and/or macroinvertebrate communities. The endpoints used to quantify macroinvertebrate response have been similar to those employed in field surveys to detect xenobiotic impacts. Two early studies determined temperature[10] and surfactant[11] tolerances of gnotobiotic assemblages of aquatic insects. Most recent studies have quantified effects on macroinvertebrate community structure by reporting changes in species richness,[17,25,29] total numbers,[17,25,28,29] diversity indices,[17,25,28,29] or densities of individual taxa.[25,28,32,39] All of these endpoints have proven sensitive to stressors, and most are ecologically meaningful. Decisions on how to interpret and report multispecies test results depend on what questions are being asked. However, analysis of effects on community structure should begin by determining effects on individual species. In this manner, the relative sensitivities of component species, and population and community-level NOECs can be determined. A number of statistical procedures have been employed as summary measures of effects on community structure;[cf.93] those methods are addressed elsewhere in this book.

Some studies with macroinvertebrate communities have included measurements of change in functional attributes such as drift,[17,22,29,32] secondary production,[17] predator–prey interactions,[26,94] biomass,[17,22,28,29] functional feeding groups,[25,28,39] and community respiration.[29,16] Of these attributes, macroinvertebrate drift is probably the most sensitive measure of community stress. At a minimum, drift should be quantified in the first hours following initial exposure. Dramatic increases in macroinvertebrate drift in response to treatment[17,29] could significantly affect community structure in stream microcosms that do not prevent emigration. Although drift is the primary stressor avoidance mechanism for macroinvertebrates in natural systems, the absence of a drift response in stream microcosm studies does not preclude significant effects on community structure.[32] Most functional processes controlled by lotic macroinvertebrates are dependent upon community structure. However, because of functional redundancies within most macroinvertebrate communities, stressor-induced changes in community structure will typically occur at toxicant concentrations lower than those that affect ecosystem function. Therefore, with respect to macroinvertebrates, change in community structure is usually the most sensitive indicator of toxicant stress. Although such studies require trained taxonomists, the composition of macroinvertebrate communities is easier to quantify than that of microbial communities.

The endpoints most commonly used to determine toxicant effects on microbial communities are changes in chlorophyll levels[21,29,33–35,37,39] and total biomass.[21,29,33,35,39] Other endpoints measured include changes in relative abundances,[21,28,33,37] total densities,[37,38] species richness,[33] production/respiration ratios,[16,29,33] and nutrient uptake rates.[35] All of the above endpoints have been shown to be sensitive to chemical stressors, but the measures of change in microbial community structure such as decreased species richness and shifts in relative abundance are generally more sensitive than changes in functional processes.[95] The expertise and time required to detect stressor-induced changes in microbial community composition are considerable, and depending on test objectives and the type of stressor being tested, other measures such as changes in chlorophyll levels may be suitable response variables.

All of the responses outlined above can be analyzed by various inferential statistics to detect significant stressor-induced change. Studies that attempted field validation of laboratory predictions[17,31,32] have shown that stream microcosm tests can accurately predict both the degree and type of change in community composition to be expected from chemical exposures in natural streams. Such predictions cannot be based on the results of single species tests. Statistical techniques, field validation, and predictive utility are treated elsewhere in this volume.

CONCLUSIONS AND RECOMMENDATIONS

During the past decade, there has been a marked increase in the use of stream microcosms and mesocosms to determine the fate and effects of contaminants. Artificial stream research is no longer just an academic curiosity. The predictive utility of these test systems is now widely recognized, and some industries are beginning to incorporate results from artificial stream studies into hazard assessments for effluents and products. In this author's opinion, which may be biased by past experience, small, partially recirculating stream microcosms containing natural macroinvertebrate and microbial assemblages offer the best promise for incorporation into the regulatory framework.

Although partially recirculating stream microcosms may not, in theory, be the best tools for predicting the ultimate fate of test chemicals, they can be used to determine bioaccumulation patterns among various interacting trophic groups and are especially well suited for determining effects on macroinvertebrate and microbial communities. Partially recirculating stream microcosms are sensitive predictors of community level responses and, relative to larger microcosms and mesocosms, are more easily replicated and lower in cost. In addition, response variability within replicates is lower than that of larger test systems. However, a considerable amount of additional research is necessary in several areas.

Artificial substrates colonized in selected natural streams within a given ecoregion are probably the best source of both macroinvertebrate and microbial test communities. Variability among colonized substrates is low and easily quantified. Artificial substrates are also easy to sample during testing. Some additional research is needed to ensure that these test communities accurately reflect the composition of the natural source community. More research is also necessary to determine the range of acceptable levels for various abiotic factors during testing. At the present time, it is suggested that abiotic conditions during testing should be similar to those in the source ecosystem. Further work will also be necessary to determine the ability of these test systems to produce similar results in different laboratories and the extent to which test system responses reflect responses in the field. However, because of differences in community composition and abiotic conditions, test results from different ecoregions and dissimilar source streams within an ecoregion may vary considerably. This variability should be viewed not as an obstacle to incorporating stream microcosm testing into the regulatory framework but as an indicator of the predictive utility of these tests. Although much progress has been made in the recent past, there are still many stimulating and challenging questions in the area of artificial stream research that need to be answered.

REFERENCES

1. Tidy, C. M., River Water, *J. Chem. Soc.* 37:268–327 (1880).
2. Thomas, E. A., Versuche uber die Selbstreinigung Fliessenden Wassers, *Mitt. Geb. Lebensmittelunters. Hyg.* 35:199–218 (1944).
3. Wuhrmann, K., Uber die Biologische Prufung von Abwasserreinigungsanlagen, *Gesund. Ing.* 72:253–261 (1951).

4. Wuhrmann, K., High Rate Activated Sludge Treatment and its Relation to Stream Sanitation. II Biological River Tests of Plant Effluents, *Sewage Ind. Wastes* 26:212–220 (1954).

5. Gjullin, C. M., O. B. Cope, B. F. Quisenberry, and F. R. DuChanois, The Effect of Some Insecticides on Black Fly Larvae in Alaskan Streams, *J. Econ. Entomol.* 42(1):100–105 (1949).

6. Warren, C. E. and G. E. Davis, Laboratory Stream Research: Objectives, Possibilities, and Constraints, *Ann. Rev. Ecol. Syst.* 2:111–144 (1971).

7. Whurmann, K., Some Problems and Perspectives in Applied Limnology, *Mitt. Int. Vere. Limnol.* 20:324–402 (1974).

8. McIntire, C. D., Periphyton Assemblages in Laboratory Streams, in *River Ecology*, B. A. Whitton, Ed., Berkeley, CA: University of California Press (1975), pp. 403–430.

9. Phinney, H. K. and C. D. McIntire, Effect of Temperature on Metabolism of Periphyton Communities Developed in Laboratory Streams, *Limnol. Oceanogr.* 10:341–44 (1965).

10. Nebeker, A. V. and A. E. Lemke, Preliminary Studies on the Tolerance of Aquatic Insects to Heated Waters, *J. Kansas Entomol. Soc.* 41:413–418 (1968).

11. Surber, E. W. and T. O. Thatcher, Laboratory Studies of the Effects of Alkyl Benzene Sulfonate (ABS) on Aquatic Invertebrates, *Trans. Am. Fish. Soc.* 92:152–160 (1963).

12. Wilton, D. P. and B.V. Travis, An Improved Method for Simulated Stream Tests of Blackfly Larvacides, *Mosq. News* 25:118–123 (1965).

13. Rose, F. L. and D. McIntire, Accumulation of Dieldrin by Benthic Algae in Laboratory Streams, *Hydrobiologia* 35:481–493 (1970).

14. Biological Field and Laboratory Methods for Measuring the Quality of Surface Waters and Effluents, National Environmental Research Center, U. S. EPA Report-670/4–73–001 (1973).

15. Maki, A. W., L. W. Geissel, and H. E. Johnson, Comparative Toxicity of Larval Lampricide (TFM:3-trifluoromethyl-4-nitrophenol) to Selected Benthic Macroinvertebrates, *J. Fish. Res. Board Can.* 32:1455–1459 (1975).

16. Maki, A. W. and H. E. Johnson, Evaluation of a Toxicant on the Metabolism of Model Stream Communities, *J. Fish. Res. Board Can.* 33:2740–2746 (1976).

17. Maki, A. W., Evaluation of Toxicant Effects on Structure and Function of Model Stream Communities: Correlations with Natural Stream Effects, in *Microcosms in Ecological Research*, J. P. Giesy, Jr., Ed., Springfield, VA: National Technical Information Service (1980), pp. 583–609.

18. Burks, S. L. and J. L. Wilhm, Bioassays with a Natural Assemblage of Benthic Macroinvertebrates, in *Aquatic Toxicology and Hazard Evaluation*, F.L. Mayer and J. L. Hamelink, Eds., Philadelphia, PA: ASTM STP 634, American Society for Testing and Materials (1977), pp. 127–136.

19. Sigmon, C. F., H. F. Kania, and R. J. Beyers, Reductions in Biomass and Diversity Resulting from Exposure to Mercury in Artificial Streams, *J. Fish. Res. Board. Can.* 34:493–500 (1977).

20. Giesy, J. P., Jr., H. J. Kania, J. W. Bowling, R. L. Knight, S. Mashburn, and S. Clarkin, Fate and Biological Effects of Cadmium Introduced into Channel Microcosms, Environmental Protection Agency, Athens, GA., U. S. EPA Report-6002/3–79–039 (1979).

21. Gerhart, D. Z., S. M. Anderson, and J. Richter, Toxicity Bioassays with Periphyton Communities: Design of Experimental Streams, *Water Res.* 11:567–570 (1977).

22. Seim, W. K., J. A. Lichatowich, R. H. Ellis, and G. E. Davis, Effects of Kraft Mill Effluents on Juvenile Salmon Production in Laboratory Streams, *Water Res.* 11:189–196 (1977).

23. Ohtake, H., S. Aiba, and R. Sudo, Growth and Detachment of Periphyton in an Effluent from the Secondary Treatment Plant of Wastewaters, *Jpn. J. Limnol.* 39(4):163–169 (1978).

24. Giesy, J. P., Ed., *Microcosms in Ecological Research,* Springfield,VA: National Technical Information Service (1980), 1110 pp.

25. Clements, W. H., D. S. Cherry, and J. Cairns, Jr., Structural alterations in Aquatic Insect Communities Exposed to Copper in Laboratory Streams, *Environ. Toxicol. Chem.* 7:715–722 (1988).

26. Clements, W. H., D. S. Cherry, and J. Cairns, Jr., The Influence of Copper Exposure on Predatory–Prey Interactions in Aquatic Insect Communities, *Freshwater Biol.* 21:483–488 (1989).

27. Coats, J. R., A Stream Microcosm for Environmental Assessment of Pesticides, in *Microcosms in Ecological Research*, J. P. Giesy, Jr., Ed., Springfield, VA: National Technical Information Service (1980), pp. 715–723.

28. Hansen, S. R. and R. R. Garton, Ability of Standard Toxicity Tests to Predict the Effects of the Insecticide Diflubenzuron on Laboratory Stream Communities, *Can. J. Fish. Aquat. Sci.* 39:1273–1288 (1982).

29. Lynch, T. R., H. E. Johnson, and W. J. Adams, Impact of Atrazine and Hexachlorobiphenyl on the Structure and Function of Model Stream Ecosystems, *Environ. Toxicol. Chem.* 4:399–413 (1985).

30. Pontasch, K. W., and J. Cairns, Jr. Establishing and Maintaining Laboratory-Based Microcosms of Riffle Insect Communities: Their Potential for Multispecies Toxicity Tests, *Hydrobiologia* 175:49–60 (1989).

31. Pontasch, K. W., B. R. Niederlehner, and J. Cairns, Jr., Comparisons of Single-Species, Microcosm, and Field Responses to a Complex Effluent, *Environ. Toxicol. Chem.* 8:521–532 (1989).

32. Pontasch, K. W. and J. Cairns, Jr., Multispecies Toxicity Tests Using Indigenous Organisms: Predicting the Effects of Complex Effluents in Streams, *Arch. Environ. Contam. Toxicol.* 20:103–112 (1991).

33. Hamala, J. A. and H. P. Kollig, The Effects of Atrazine on Periphyton Communities in Controlled Laboratory Ecosystems, *Chemosphere* 14(9):1391–1408 (1985).

34. Krewer, J. A. and H. W. Holm, The Phosphorus-Chlorophyll a Relationship in Periphytic Communities in a Controlled Ecosystem, *Hydrobiologia* 94:173–176 (1982).

35. Krieger, K. A., D. B. Baker, and J. W. Kramer, Effects of Herbicides of Stream Aufwuchs Productivity and Nutrient Uptake, *Arch. Environ. Contam. Toxicol.* 17:299–306 (1988).

36. Lynch, T. R., H. E. Johnson, and W. J. Adams, The Fate of Atrazine Hexachlorobiphenyl Isomer in Naturally-Derived Model Stream Ecosystems, *Environ. Toxicol. Chem.* 1:179–192 (1982).

37. Maurice, C. G., R. L. Lowe, T. M. Burton, and R. M. Stanford, Biomass and Compositional Changes in the Periphytic Community of an Artificial Stream in Response to Lowered pH, *Water, Air, and Soil Pollut.* 33:165–177 (1987).

38. Schwab, B. S., D. A. Maruscik, R. M. Ventullo, and A. C. Palmisano, Adaptation of Periphytic Communities in Model Streams to a Quaternary Ammonium Surfactant, *Environ. Toxicol. Chem.* 11:1169–1177 (1992).

39. Selby, D. A., J. M. Ihnat, and J. J. Messer, Effects of Subacute Cadmium Exposure on a Hardwater Mountain Stream Microcosm, *Water Res.* 19(5):645–655 (1985).

40. Shimp, R. J., B. S. Schwab, and R. J. Larson, Adaptation to a Quaternary Ammonium Surfactant by Suspended Microbial Communities in a Model Stream, *Environ. Toxicol. Chem.* 8:723–730 (1989).

41. Allard, M. and G. Moreau, Effects of Experimental Acidification on a Lotic Macroinvertebrate Community, *Hydrobiologia* 144:37–49 (1987).

42. Arthur, J. W., J. A. Zischke, and G. L. Ericksen, Effect of Elevated Water Temperature on Macroinvertebrate Communities in Outdoor Experimental Channels, *Water Res.* 16:1465–1477 (1982).

43. Arthur, J. W., J. A. Zischke, K. N. Allen, and R. O. Hermanutz, Effects of Diazinon on Macroinvertebrates and Insect Emergence in Outdoor Experimental Channels, *Aquat. Toxicol.* 4:283–301 (1983).

44. Clements, W. H., D. S. Cherry, and J. Cairns, Jr., Impact of Heavy Metals on Insect Communities in Streams: a Comparison of Observational and Experimental Results, *Can. J. Fish. Aquat. Sci.* 45:2017–2025 (1988).

45. Clements, W. H., J. L. Farris, D. S. Cherry, and J. Cairns, Jr., The Influence of Water Quality on Macroinvertebrate Community Responses to Copper in Outdoor Experimental Channels, *Aquat. Toxicol.* 14:249–262 (1989).

46. Crossland, N. O., G. C. Mitchell, and P. B. Dorn, Use of Outdoor Artificial Streams to Determine Threshold Toxicity Concentrations for a Petrochemical Effluent, *Environ. Toxicol. Chem.* 11:49–59 (1992).

47. Hall, T. J., R. K. Haley, and L. E. LaFleur, Effects of Biologically Treated Bleached Kraft Mill Effluent on Cold Water Stream Productivity in Experimental Stream Channels, *Environ. Toxicol. Chem.* 10:1051–1060 (1991).

48. Kreutzweiser, D. P. and S. S. Capell, A Simple Stream-Side Test System for Determining Acute Lethal and Behavioral Effects of Pesticides on Aquatic Insects, *Environ. Toxicol. Chem.* 11:993–999 (1992).

49. Newman, R. M., J. A. Perry, E. Tam, and R. L. Crawford, Effects of Chronic Chlorine Exposure on Litter Processing in Outdoor Experimental Streams, *Freshwater Biol.* 18:415–528 (1987).

50. Nordlie, K. J. and J. W. Arthur, Effect of Elevated Water Temperature on Insect Emergence in Outdoor Experimental Channels, *Environ. Pollut. (series A)* 25:53–65 (1981).

51. Perkins, J. L., Bioassay Evaluation of Diversity and Community Comparison Indexes, *J. Water Pollut. Control Fed.* 55(5):522–530 (1983).

52. Rodgers, E. B., Effects of Elevated Temperatures on Macroinvertebrate Populations in the Browns Ferry Experimental Ecosystems, in *Microcosms in Ecological Research*, J. P. Giesy, Jr., Ed., Springfield, VA: National Technical Information Service (1980), pp. 684–702.

53. Stout, R. J. and W. E. Cooper, Effect of p-Cresol on Leaf Decomposition and Invertebrate Colonization in Experimental Outdoor Streams, *Can. J. Fish. Aquat. Sci.* 40:1647–1657 (1983).

54. Zischke, J. A., J. W. Arthur, K. J. Nordlie, R. O. Hermanutz, D. A. Standen, and T. P. Henry, Acidification Effects on Macroinvertebrates and Fathead Minnows (*Pimephales promelas*) in Outdoor Experimental Channels, *Water Res.* 17:47–63 (1983).

55. Armitage, B. J., Effects of Temperature on Periphyton Biomass and Community Composition in the Browns Ferry Experimental Channels, in *Microcosms in Ecological Research*, J. P. Giesy, Jr., Ed., Springfield, VA: National Technical Information Service (1980), pp. 668–683.

56. Colwell, F. S., S. G. Hornor, and D. S. Cherry, Evidence of Structural and Functional Adaptation in Epilithon Exposed to Zinc, *Hydrobiologia* 171:79–90 (1989).

57. Dean-Ross, D., Response of Attached Bacteria to Zinc in Artificial Streams, *Can. J. Microbiol.* 36:561–566 (1990).

58. Genter, R. B., D. S. Cherry, E. P. Smith, and J. Cairns, Jr., Attached-Algal Abundance Altered by Individual and Combined Treatments of Zinc and pH, *Environ. Toxicol. Chem.* 7:723–733 (1988).

59. Genter, R. B., D. S. Cherry, E. P. Smith, and J. Cairns, Jr., Algal-Periphyton Population and Community Changes from Zinc Stress in Stream Mesocosms, *Hydrobiologia* 153:261–275 (1987).

60. Hoffman, R. W. and A. J. Horne, On-Site Flume Studies for Assessment of Effluent Impacts on Stream Aufwuchs Communities, in *Microcosms in Ecological Research*, J. P. Giesy, Jr., Ed., Springfield, VA: National Technical Information Service (1980), pp. 610–624.

61. Kosinski, R. J., The Effect of Terrestrial Herbicides on the Community Structure of Stream Periphyton, *Environ. Pollut. (Series A)* 36:165–189 (1984).

62. Kosinski, R. J. and M. G. Merkle, The Effect of Four Terrestrial Herbicides on the Productivity of Artificial Stream Algal Communities, *J. Environ. Qual.* 13:75–82 (1984).

63. Manuel, C. Y. and G. W. Minshall, Limitations on the Use of Microcosms for Predicting Algal Response to Nutrient Enrichment in Lotic Systems, in *Microcosms in Ecological Research*, J. P. Giesy, Jr., Ed., Springfield, VA: National Technical Information Service (1980), pp. 645–667.

64. Moorhead, D. L. and R. J. Kosinski, Effect of Atrazine on the Productivity of Artificial Stream Algal Communities, *Bull. Environ. Contam. Toxicol.* 37:330–336 (1986).

65. Pignatello, J. J., M. M. Martinson, J. G. Steiert, R. E. Carlson, and R. L. Crawford, Biodegradation and Photolysis of Pentachlorophenol in Artificial Freshwater Streams, *Appl. Environ. Microbiol.* 45(5):1024–1031 (1983).

66. Pignatello, J. J., L. K. Johnson, M. M. Martinson, R. E. Carlson, and R. L. Crawford, Response of the Microflora in Outdoor Experimental Streams to Pentachlorophenol: Compartmental Contributions, *Appl. Environ. Microbiol.* 50(1):127–132 (1985).

67. Pignatello, J. J., L. K. Johnson, M. M. Martinson, R. E. Carlson, and R. L. Crawford, Response of the Microflora in Outdoor Experimental Streams to Pentachlorophenol: Environmental Factors, *Can. J. Microbiol.* 32:38–46 (1986).

68. Pratt, J. R. and N. J. Bowers, Effect of Selenium on Microbial Communities in Laboratory Microcosms and Outdoor Streams, *Toxicity Assess.* 5:293–307 (1990).

69. Rodgers, J. H., Jr., J. R. Clark, K. L. Dickson, and J. Cairns, Jr., Nontaxonomic Analyses of Structure and Function of Aufwuchs Communities in Lotic Microcosms, in *Microcosms in Ecological Research*, J. P. Giesy, Jr., Ed., Springfield, VA: National Technical Information Service (1980), pp. 625–644.

70. Vymazal, J., The Use of Periphyton Communities for Nutrient Removal from Polluted Streams, *Hydrobiologia* 166:225–237 (1988).

71. Testing for Effects of Chemicals on Ecosystems, National Research Council, National Academy Press, Washington, DC (1981).

72. Cairns, J., Jr., Are Single Species Tests Alone Adequate for Estimating Hazard, *Hydrobiologia* 137:271–278 (1983).

73. Kimball, K. D. and S. A. Levin, Limitations of Laboratory Bioassays: The Need for Ecosystem Level Testing, *BioScience* 35:165–171 (1985).

74. Touart, L. W., Aquatic Mesocosm Test to Support Pesticide Registrations, U.S. EPA Report-540/09–88–035 (1988).

75. Workshop on Aquatic Microcosms for Ecological Assessment of Pesticides, Pensacola, FL: Society of Environmental Toxicology and Chemistry (1992), 56 pp.

76. Odum, E. P., The Mesocosm, *BioScience* 34(9):558–562 (1984).

77. Shriner, C. and T. Gregory, Use of Artificial Streams for Toxicological Research, *CRC Crit. Rev. Toxicol.* 13(3):253–281 (1984).

78. Clubb, R. W., A. R. Gaufin, and J. L. Lord, Acute Cadmium Toxicity Tests upon Nine Species of Aquatic Insects, *Environ. Res.* 9:332–341 (1975).

79. Spehar, R. L., R. L. Anderson, and J. T. Fiandt, Toxicity and Bioaccumulation of Cadmium and Lead in Aquatic Invertebrates, *Environ. Pollut.* 15:195–208 (1978).

80. Jensen, L. D. and A. R. Gaufin, Effects of Ten Organic Insecticides on Two Species of Stonefly Naiads, *Trans. Am. Fish. Soc.* 93:27–34 (1964).

81. Sanders, H. O. and O. B. Cope, The Relative Toxicities of Several Pesticides to Naiads of Three Species of Stonefly, *Limnol. Oceanogr.* 33:112–117 (1968).

82. Gauss, J. D., P. E. Woods, R. W. Winner, and J. H. Skillings, Acute Toxicity of Copper to Three Life Stages of *Chironomus tentans* as Affected by Water Hardness-Alkalinity, *Environ. Pollut.* 37:149–157 (1985).

83. Williams, K. A., D. W. J. Green, D. Pascoe, and D. E. Gower, The Acute Toxicity of Cadmium to Different Larval Stages of *Chironomus riparius* (Diptera: Chironomidae) and its Ecological Significance For Pollution Regulation, *Oecologia* 70:362–366 (1986).

84. Sweeney, B. W., Factors Influencing Life-History Patterns of Aquatic Insects, in *The Ecology of Aquatic Insects*, V. H. Resh and D. M. Rosenberg, Eds., New York, NY: Praeger Publishers, Inc. (1984), pp. 56–100.

85. Hammons, A. S., Ed., Ecotoxicological Test Systems. Proceedings of a Series of Workshops, Office of Toxic Substances, U.S. EPA Report-560/6–81–004 (1981).

86. McIntire, C. D., Some Effects of Current Velocity on Periphyton Communities in Laboratory Streams, *Hydrobiologia* 27:559–570 (1966).

87. Peterson, C. J. and R. J. Stevenson, Substratum Conditioning and Diatom Colonization in Different Current Regimes, *J. Phycol.* 25:790–793 (1989).

88. Steinman, A. D. and C. D. McIntire, Effects of Current Velocity and Light Energy on the Structure of Periphyton Assemblages in Laboratory Streams, *J. Phycol.* 22:352–361 (1986).

89. Steinman, A. D., P. J. Mulholland, and D. B. Kirschtel, Interactive Effects of Nutrient Reduction and Herbivory on Biomass, Taxonomic Structure, and P Uptake in Lotic Periphyton Communities, *Can. J. Fish. Aquat. Sci.* 48(10):1951–1959 (1991).

90. Steinman, A. D., P. J. Mulholland, A. V. Palumbo, T. F. Flum, and D. L. DeAngelis, Resilience of Lotic Ecosystems to a Light-Elimination Disturbance, *Ecology* 72(4):1299–1313 (1991).

91. Clements, W. H., J. H. Van Hassel, D. S. Cherry, and J. Cairns, Jr., Colonization, Variability, and the Use of Substratum-Filled Trays for Biomonitoring Benthic Communities, *Hydrobiologia* 173:45–53 (1988).

92. Rosenberg, D. M. and V. H. Resh, The Use of Artificial Substrates in the Study of Freshwater Benthic Macroinvertebrates, in *Artificial Substrates*, J. Cairns, Jr., Ed., Ann Arbor, MI: Ann Arbor Sci. Pub. (1982), pp. 175–235.

93. Pontasch, K. W., E. P. Smith, and J. Cairns, Jr., Diversity Indices, Community Comparison Indices and Canonical Discriminant Analysis: Interpreting The Results Of Multispecies Toxicity Tests, *Water Res.* 23(10):1229–1238 (1989).

94. Pontasch, K. W., Predation by *Paragnetina fumosa* (Banks) (Plecoptera:Perlidae) on Mayflies: The Influence of Substrate Complexity, *Am. Midl. Nat.* 119(2):441–443 (1988).

95. Pratt, J. R., Aquatic Community Response to Stress: Prediction and Detection Of Adverse Effects, in *Aquatic Toxicology and Risk Assessment: Thirteenth Volume,* ASTM STP 1096, W. G. Landis and W. H. van der Schalie, Eds., Philadelphia, PA: American Society for Testing and Materials (1990), pp. 16–26.

CHAPTER 11

Landscape-Scale Effects of Toxic Events for Ecological Risk Assessment

Lenore Fahrig and Kathryn Freemark

INTRODUCTION

The effect of spatial heterogeneity on response of organisms has rarely been considered in ecological risk assessment or in ecotoxicological studies (Emlen 1989, Jorgensen 1990). Barnthouse (1992) cites examples of regional or "landscape" models of the distribution of ecosystems or land-use types to quantify ecological effects of regionally distributed stresses such as acid deposition, ozone, and climate change. In addition, Barnthouse (1992) recognizes the application of agricultural pesticides as a potentially important regional problem. Increased and widespread use of agricultural pesticides has been linked with population declines of plant, invertebrate, and vertebrate species (Grue et al. 1986, O'Connor and Shrubb 1986, Streibig 1988, Sheehan et al. 1987, Fox et al. 1989, Fuller et al. 1991, Sotherton 1991, Tew et al. 1992, Freemark and Boutin, in press).

In this chapter we argue that ecological risk assessments for toxicants must go beyond the current practice of only assessing effects at the spatial and temporal scales of direct impact (Cairns 1984). Our objectives are (1) to extend results of the effects of landscape pattern and variability on population survival to landscape-scale effects of toxic events, (2) to develop a conceptual model of landscape-scale effects of toxic events on species, and (3) to propose a practical approach for predicting when landscape-scale effects of toxic events are likely to occur. To date there has been little if any research on the interacting effects of landscape pattern and toxic events for ecological risk assessment. Therefore, we rely primarily on studies from the general ecological literature, particularly those on the dynamics of stochastic, spatially structured populations. We do so by viewing a toxic event as just another type of ecological disturbance, thus enabling us to argue by extension from the existing literature (e.g., Fahrig 1990).

LANDSCAPE-SCALE PROPAGATION OF LOCAL EXTINCTIONS

Many, if not all, natural populations are comprised of many local populations. At the local scale, naturally occurring extinctions are frequent. Factors that cause population reductions such as disease, predation, and inclement weather can reduce a population to low levels, making local extinction likely (Paine 1988, Berger 1990). In many cases, "demographic stochasticity" results in the final extirpation (Hanski 1986). Demographic stochasticity includes factors such as skewed sex ratios or age ratios, which occur by chance. For example, if all five individuals surviving a disease happen to be male or past reproductive age, the population will die out.

High rates of natural (i.e., not caused by a toxic event) local extinctions have been documented. Addicott (1978) showed that local populations of aphids on fireweeds became extinct and colonizations occurred throughout the summer. In 23 wooded patches of 1 to 24 ha, Merriam and Wegner (1992) found between 8.7 and 15.2% of patches per year suffered local extinctions of white-footed mouse (*Peromyscus leucopus*). In a 5-year study of ambush bugs (*Phymata americana*) living in goldenrod (*Solidago*) patches, Mason (1977) found annual local extinction rates of 0 to 10% of 10 to 23 local populations. Van der Meijden et al. (1985) found that between 9 and 64% of local populations of nine species of biennial dune plants suffered local extinctions due to severe drought. Paine (1988) estimated the local extinction rate of sea palms to be between 0 and 64% per year. Finally, a number of studies have reported significant extinction rates even in highly vagile birds (Freemark 1989, Villard et al. 1992, Verboom et al. 1991).

Although local extinctions are common, the population at the larger regional or landscape scale (i.e., the metapopulation) can be stable. For example, populations of white-footed mice (*Peromyscus leucopus*) in small woodlots are highly variable, but, when integrated over several woodlots, the metapopulaton is more stable (Merriam and Wegner 1992). Ehrlich et al. (1975) and Thomas (1991) showed that population sizes of the Bay checkerspot butterfly (*Euphydryas editha bayensis*) and the silver-studded blue butterfly (*Plebejus argus*), respectively, are less variable at a large spatial scale than at a small scale. Villard et al. (1992) observed year-to-year dynamics in habitat patch occupancy at the landscape scale for three species of migratory birds breeding in temperate forests. The overall picture is of unstable population dynamics at the local scale but persistence of the metapopulations at the landscape scale because local extinctions are balanced by recolonizations.

This scenario of natural spatiotemporal fluctuations has implications for the impact of toxic events on population survival. If a toxic event wipes out a local population that is a critical source of colonists for other local populations (that may have experienced local extinctions unrelated to the toxic event), these other areas will not be recolonized. Other local areas that depend on these for colonists will also not be colonized, and so on. In this way the population reduction from a single local toxic event can be propagated through the landscape. In extreme cases, instability in metapopulation dynamics can result, causing regional extinction. Therefore, the effect of a toxic event can extend beyond the area of direct impact and can last much longer than the duration of effectiveness of the substance. In the remainder of this chapter we refer to this phenomenon as "landscape-scale propagation" of toxic event(s).

THE CONSTRAINT OF LANDSCAPE STRUCTURE

The landscape-scale outcome of a toxic event can range from no detectable effect, to a temporary reduction in landscape population size, to the extreme of regional extinction. The actual effect depends on (1) the spatiotemporal con-figuration of the population in the landscape, including between-area differences in population variability and habitat size, quality, and accessibility (Kareiva 1990) and (2) the spatiotemporal distribution of toxic event(s) in the same landscape (Fahrig 1990). In most cases, the spatial distribution of the population in the landscape is better defined by its "potential" spatial distribution since, at any time, some areas that could serve as habitat are empty due to local extinctions. The spatial distribution is therefore best described by the spatial distribution of habitat rather than the population itself. There are two general kinds of habitat distributions: patchy and continuous. The patchy population is subdivided into discrete local populations in suitable "habitat patches" surrounded by a "matrix" of unsuitable habitat, while the habitat of the continuous population is interconnected.

Populations that are continuous (habitat is interconnected at the landscape scale) are resistant to local extinctions (Fahrig 1991). This is because dispersal to the locally extinct area is not restricted by the spatial configuration of the habitat. Recolonization of the area occurs quickly from the neighboring habitat. However, many, if not most, populations are patchily distributed. In this case the large-scale implications of a local extinction depend on the spatial relation-ships between the affected area and other patches of habitat. To predict the effect, it is necessary to know whether the area in which the local extinction occurred is critical to metapopulation survival. The area is critical to metapopulation survival if it provides immigrants to other areas that would otherwise not sustain local populations (Pulliam 1988, Howe et al. 1991) and/or it provides a "stepping stone" (Kareiva 1990) between pairs of patches, thus allowing movement between them for recolonizations of local extinctions. The status of the affected area (critical or not) depends on the spatial structure of the landscape (Fahrig and Merriam, 1994).

Landscape structure is defined as the extent and composition of habitat elements and their spatial arrangement (Dunning et al. 1992). Habitat elements are defined by the way in which they are used by a particular species. Examples are patches of breeding habitat or feeding habitat, areas of inhospitable habitat, and areas that can be used as dispersal corridors. Important characteristics of landscape structure include the following:

1. **Size, shape, and quality of patches** — the size of a patch has been shown to influence local population persistence. Verboom et al. (1991) and Villard et al. (1992) found a positive relationship between habitat patch size and persistence of local populations of forest birds. Also, positive relationships between patch size and local population size (e.g., Lynch and Whigham 1984) and between local population size and population persistence (e.g., Paine 1988, Berger 1990) imply a positive relationship between patch size and population persistence.

   Patch shape can influence population survival if the population dynamics change in relation to the distance from the edge of the patch. For example, Gates and Gysel (1978) found that the number of passerine birds increased at the field-forest edge. Ambuel and Temple (1983) suggest that forest-edge and farmland bird species exclude certain forest-dwelling species and that this exclusion has a greater impact than changes in patch area or isolation. Best et al. (1990) found that bird species abundance decreased as the size of crop fields increased because of a reduced edge effect. They predicted that narrower and/or more irregularly shaped fields would support more birds because they have proportionately more edge. Thus, two patches of the same area but having different amounts of edge may have different population dynamics.

   Pulliam (1988) simplified the issue of patch quality by referring to two qualitatively different kinds of patches: sources and sinks. Source patches are those that, on average, produce more offspring than they are able to maintain. The excess emigrate and are potential colonists of other patches. In sink patches, the habitat quality is low, so the productivity rate is below that needed to sustain a population over the long term. Sink patches are therefore dependent on source patches for their persistence. Best (1986) argues that croplands can act as sink habitats for many bird species because reproduction and survival are reduced by frequent and untimely agricultural disturbances including pesticide applications. Although sink populations are not self-sustaining, Howe et al. (1991) have demonstrated that they can make a significant contribution to the size and longevity of the metapopulation. In addition to creating sink habitats, pesticides, and presumably other toxicants, can also lower the carrying capacity of habitats (Sheehan et al. 1987, Sotherton 1991, Fuller et al. 1991, Rogers and Freemark 1991). This results in smaller populations at the local and landscape scales, which increases the probability of local extinction from stochastic processes and decreases rates of recolonization from reductions in numbers of dispersing individuals. This outcome is particularly likely when the toxic event is not of short duration but is chronic (e.g., acid deposition).

2. **The presence of dispersal routes through the landscape** — Fahrig and Merriam (1985) showed that fencerows, which function as dispersal routes for white-footed mice (*Peromyscus leucopus*) in a landscape of woodlots and farmland, are important for regional population abundance. Dmowski and Kozakiewicz (1990) found that a shrub strip between a forest and a littoral zone acted as a corridor enhancing movements of birds. Bennett (1990) showed that survival of regional populations of small mammals in a fragmented landscape in Australia was facilitated by movements between remnant forest patches along vegetated corridors (but see Simberloff et al. 1992). Verboom and van Apeldoorn (1990) found that the amount of hedgerow surrounding wooded fragments was positively related to the presence of red squirrel (*Sciuris vulgaris*) in the fragment. Clearly, habitat patches themselves can also act as dispersal routes within the landscape.

3. **The quality of dispersal routes** — quality can affect the likelihood of dispersers using the route and/or the probability that dispersers using the route will survive. Henein and Merriam (1990) showed in a simulation study that the quality of dispersal routes can be important for the abundance of the regional population. Availability of corridors with qualities permitting both breeding and movement between patches is also important to population survival (Bennett 1987). Extreme low quality is equivalent to a barrier effect. For example, Duelli (1990) showed that edges of agricultural fields can be barriers to movement through the landscape by some farmland arthropods. Mader et al. (1990) showed that minor roads blocked the dispersal of some forest arthropods.

4. **Spatial configuration of habitat elements** — Lefkovitch and Fahrig (1985) showed in a simulation study that, for population survival and abundance, the total number of dispersal corridors in the landscape may be less important than their configuration relative to the habitat patches. In particular, they found that it is the overall shape and size of the geometric figure formed by interconnected patches that is most important; large, closed figures produce the highest probability of regional persistence. Seno (1988) modeled a one-dimensional patchy population, in which one patch was different from the others in terms of population demography and/or dispersal. The model showed that the spatial location of this patch affected the regional population

dynamics; the more central the patch, the greater its impact. Herben et al. (1991) showed in a model of moss population dynamics that degree of clumping of substrate patches affects population persistence. The importance of clumping for increasing persistence has also been demonstrated in modeling studies (Adler and Nürnberger 1994).

Many field studies have demonstrated the importance of spatial configuration of the landscape, particularly the importance of patch isolation on local population abundance and persistence. Lawton and Woodroffe (1991) showed that breeding water voles were less likely to be present in isolated sites. Dingle (1991) found that the large milkweed bug was less likely to be found in host patches that were far from major rivers. He hypothesized that this was due to the fact that the bugs follow water courses during migration. Potter (1990) found that degree of isolation of forest remnants affected the probability of use by the brown kiwi in New Zealand. If large remnants are interspersed with small ones, the kiwis can move between the large remnants by using the small ones as "stepping stones". Burel (1989) reported similar use of stepping stones by forest carabid beetles in Brittany farmland. Freemark and Collins (1992) have shown that the variety and abundance of forest birds in small forest patches are positively related to the amount of forest nearby and proximity to larger forests.

IMPORTANCE OF DISPERSAL AND HABITAT-SPECIFIC DEMOGRAPHICS

Dispersal acts to increase population stability by decreasing the effects of disturbances (or toxic events) (Reddingius and den Boer 1970, Roff 1974a, Vance 1980), except for the rare situations when dispersal rate is highest at low population densities (Vance 1984). Our use of "dispersal" refers to a change in location of organisms, accompanied by reproduction. Reproduction can occur after the move to the new area, or the organisms may reproduce in the original area, and their offspring disperse to a new one (Lidicker 1975). The effect of landscape spatial structure on movement and recolonization depends on the dispersal characteristics of the organism.

The components of dispersal that are important for population response to the constraint of landscape structure are as follows (Fahrig and Paloheimo 1988a, Fahrig 1991):

1. Dispersal probability, or the probability of individuals leaving a patch per unit time.
2. Dispersal distance, or the probability of individuals successfully reaching a range of distances or landscape components. If the dispersers do not travel far enough to move among habitat patches, they will not be able to recolonize local extinctions.

3. Temporal pattern of dispersal. For example, dispersal may be seasonal. If local extinctions (or toxic events) are more likely to occur at some times of the year than others, the timing of movement that results in recolonization will affect its demographic impact.
4. Dispersal behavior that improves the probability of dispersers finding habitat patches and/or decreases the risk of disperser mortality. For example, some herbivorous insects including bark beetles (McMullen and Atkins 1962), desert locusts (Wallace 1958), and cabbage root fly (Prokopy et al. 1983) locate and orient toward host plant patches from a distance.

Fahrig and Paloheimo (1988a) conducted a simulation experiment in which they examined the relative importance of various components of demography and dispersal on regional abundance in patchy environments. In general, dispersal was shown to be more important than demography (e.g., birth rate) in determining regional population abundance. This result is supported in a study of small rodent populations in which immigration was found to be more important than local demographics in affecting local population persistence (Blaustein 1981). Fahrig and Paloheimo (1988a) also found that the most important determinant of regional population abundance was the probability that dispersers successfully locate new patches. The exact spatial pattern of habitat patches was found to be most important when dispersal distances are small relative to the distance between patches. For highly mobile species such as birds, habitat-specific demographic factors such as survival and fecundity have a strong impact on metapopulation size and persistence (Urban and Shugart 1986, Lande 1988, Temple and Cary 1988, Noon and Biles 1990). Toxic events can have a direct effect on reproduction and survival of local populations, which, in turn, can propagate to the landscape scale (Pulliam 1994).

SCENARIOS FOR LANDSCAPE-SCALE PROPAGATION OF TOXIC EVENTS

In this section, we discuss toxic events in the following three types of exposure scenarios: (1) single local toxic event, (2) several local toxic events, and (3) large-scale toxic event. For each of these we consider effects on both continuous and patchy populations.

SINGLE LOCAL TOXIC EVENT

We first take the case of a single, unusual local toxic event such as a localized toxic spill. We would like to predict the landscape-scale effects of this event on the survival of a particular species. In other words, will the effects of the event remain local and short-term, or will they propagate through time over the landscape to destabilize the regional population?

As for the general case of local extinctions discussed above, we do not expect the effect of a local toxic event in a continuous population to propagate through the landscape. This is because dispersal to the area is not restricted by the spatial configuration of the habitat. Recolonization is expected to occur quickly from the neighboring areas. Even if the toxic event is chronic, its effects will remain local.

However, if the population is patchy, the large-scale effect of the toxic event depends on the predisturbance relationship of the population in that area to the dynamics of the regional population. In particular, it depends on whether that population was critical to regional survival. This in turn depends, as described above, on the spatial structure of the landscape, and the dispersal characteristics of the organism. In general, the larger the landscape and the more local populations, the more resistant it is to the effects of a single local toxic event (Agur and Deneubourg 1985, Roff 1974b). The longer-lasting or more chronic the local effects of the toxic event (e.g., a toxic dump site), the more likely it is to propagate to the landscape.

MANY LOCAL TOXIC EVENTS

Many toxic substances in the environment are distributed not in single rare events but as multiple, frequent events. For example, agricultural habitats experience disturbances (including applications of pesticides and other agro-chemicals) that are usually frequent, synchronous, and large scale (O'Connor and Shrubb 1986). In a continuous population in which the frequency of the toxic event is low, the above generalization for single toxic events holds: the effects of the toxic events are not expected to propagate but will remain local. However, if the frequency of toxic events is high, this generalization does not hold. In this case the toxic events themselves can restructure the population so that it is no longer continuous. The ability of organisms to recolonize affected areas may be constrained by the imposed spatial structure. Fahrig (1991) showed in simulation analyses of disturbances on a landscape, that when the frequency of disturbance (here, toxic event) is low, the dispersal distance of the organism does not affect regional population size; affected areas are recolonized even if the organism's dispersal range is small. However, if the disturbance frequency is high, there is a positive relationship between population size and dispersal distance, indicating the importance of the spatial pattern of the population in relation to the spatial pattern of the toxic events.

If either the population is patchy or the frequency of toxic events is high, the spatiotemporal pattern of the toxic events can have a large impact on the long-term survival of the population. The most important aspect of this pattern is the degree to which the toxic events in different locations are synchronous. In general, the more synchronous the events, the lower the survival probability for the regional population (Levins 1969, Harrison and Quinn 1989, Gilpin 1990, Hanski 1991).

As in the case of a single toxic event, the long-term effects of multiple events on a patchy population depend on the distribution of the events in relation to the landscape-scale dynamic of the group of local populations. To use a simple example, if all the affected areas function as source populations (Pulliam 1988, Howe et al. 1991), the stability of remaining unaffected sink areas will be decreased due to a decrease in immigration rate. If the affected areas remain uncolonized for a significant length of time, the regional population size can enter a self-perpetuated decline, which can propagate to extinction at the landscape scale. To predict the long-term effects of the toxic events it is therefore necessary to consider the interaction of these events with the extinction-recolonization dynamic on the landscape.

LARGE-SCALE TOXIC EVENT

Some toxic events, such as acid rain or ozone, occur over an area that is as large as or larger than the scale of the metapopulation. Events such as these represent the most serious threat to species survival. The immediate effect is reduction in numbers across all local populations. Some of these reductions, in combination with other stochastic factors, are likely to lead to local extinctions. Since the regional population is reduced, the probability of reestablishment of such local extinctions is reduced. If the habitat is continuous, the limiting factor is the reproductive rate of the species. If there has been a severe reduction in numbers, there will be an initial period in which chance events on the small populations may result in large-scale extinction. However, if several local populations recover to high levels, dispersal to other areas will be rapid in the continuous habitat, and the effect of the toxic event will not propagate. In a patchy population however, the regional effect of the large-scale toxic event depends on the dispersal rate, disperser survival rate, and dispersal distance of the organism affected (Fahrig and Paloheimo 1988a, Fahrig 1991) as well as the spatial structure of the landscape. In both cases, the longer lasting (more chronic) the event, the higher the probability of regional extinction.

PRACTICAL APPROACH FOR ASSESSING LANDSCAPE-SCALE EFFECTS OF TOXIC EVENTS

In the above discussion, we indicate that the landscape-scale effect of toxic event(s) on population survival depends on the landscape structure, the spatiotemporal distribution of the toxic event(s), and the dispersal characteristics and habitat-specific demographics of the organism. In other words, it depends on the interaction between the "natural" spatiotemporal functioning of the population (the extinction-recolonization dynamic) and the spatiotemporal dynamic of the toxic event(s). To make a prediction for a particular case, it will in general be necessary to develop a model that incorporates these spatiotemporal dynamics.

Since model development always has an intuitive component, it is not possible to present a perfectly standardized methodology. The choices that are made about the structure of the model ultimately depend on the question(s) the model will be used to explore (Fahrig 1991). The methodology we propose includes field studies to validate and refine the ecological realism of the simulation models for subsequent use in risk assessment. The methodology steps are as follows:

1. Build a computer model that simulates the spatial distribution of habitats and toxic events and the associated population dynamics in the landscape. This can generally be done if there is some information on the dispersal and life history characteristics of the organism.
2. Conduct sensitivity analyses to identify parameters that determine whether local toxic events propagate to regional extinction.
3. Use the model to make predictions of short-term population dynamics (e.g., 1 year) for different values of the critical parameters.
4. Test the short-term predictions with field studies.
5. Conduct long-term simulation experiments to identify circumstances in which toxic events are likely to propagate to long-term, landscape-scale impacts.

Although the following examples are not of toxic events, they contain the basic elements of this methodology. Fahrig and Merriam (1985) developed a model of white-footed mice (*Peromyscus leucopus*) in patches of wooded habitat (woodlots) using literature values for parameters. They tested the short-term prediction of the model, that population size within a season depends on habitat interconnections. Lefkovitch and Fahrig (1985) then used the model to make predictions about the long-term survival of mice in different spatial configurations of woodlots. Fahrig and Paloheimo (1987) studied the dispersal behavior of cabbage butterflies (*Pieris rapae*). They incorporated the results into a model of population dynamics and tested the short-term (1 year) predictions (Fahrig and Paloheimo 1988b). The model was then used to study the general relationships between dispersal behavior and population survival (Fahrig 1988, Fahrig and Paloheimo 1988a).

This methodology is similar to methods used in Population Viability Analysis (PVA), in which one attempts to determine the survival prospects of a population. An example of this approach is a PVA of a patchy population of an endangered plant, Furbish's lousewort (*Pedicularis furbishiae*) (Menges 1990). Menges built a stochastic simulation model of the population, which included empirically observed variation in demographic parameters and estimated rates of natural catastrophes. He used the model to predict the long-term fate of the population.

A drawback of this method is that it is limited to single species analyses. Even though the extrapolation to the landscape scale is an improvement over current approaches, one would like to be able to make statements concerning the multispecies-level effects of a toxic event (Cairns 1984). Propagation to the landscape scale depends on a wide variety of factors for which we have very little empirical data. Extrapolations and/or generalizations among species is difficult because inter- and intra-family, genus, and species variations in response to chemical intoxication are large. Given the state of existing data, any predictions of potential effects on untested species will have to be tempered by many qualifiers. In addition, propagation of toxic effects to the landscape scale depends on dispersal characteristics and habitat-specific demographics of the organism. These traits are, in general, poorly known and vary substantially among species. Multispecies predictions will therefore be difficult at best (even ignoring the complications posed by species interactions). Some researchers have begun to explore the utility of comparing life history traits of species among different geographic regions to predict faunal response to changes in landscape structure (e.g., Hansen et al. 1992, Hansen and Urban, 1992). At present it appears that the most promising approach is to conduct our proposed analysis for the species that, due to high exposure, low fecundity and/or limited dispersal abilities, are most likely to experience landscape-scale propagation.

Apart from a few exceptions, there are no general rules for predicting whether the effects of toxic event(s) will propagate across the landscape to larger scales and longer time periods than the impact of the initial event(s). We suggest that the above method is a practical approach for determining whether such propagation is likely in a particular case. However, some speculation can be made without this detailed analysis. From our arguments above, we propose the following ranking, under the assumption that the toxic events are equal in their direct and chronic effects. This ranking does not take into account details such as the organism's dispersal behavior but is intended as a starting point only. The ranking is from lowest to highest probability of propagation:

1. single local toxic event in a continuous population,
2. multiple local toxic events at a low frequency and/or low synchrony in a continuous population,
3. single local toxic event in a patchy population,
4. multiple local toxic events at a high frequency and/or high synchrony in a continuous population,
5. single large-scale toxic event in a continuous population,
6. multiple local toxic events in a patchy population,
7. single large-scale toxic event in a patchy population.

SUMMARY

In this chapter, we argue that ecological risk assessments for toxicants must go beyond the current practice of only assessing effects at the spatial and temporal scales of direct impact. Propagation of the effects of toxic event(s) beyond these scales occurs because of the interaction of the toxic event(s) with the normal landscape-scale dynamic of extinctions and recolonizations of local populations. Landscape-scale propagation of a toxic event can occur if the event wipes out or greatly reduces a local population that is a critical source of colonists for other local populations (that may have experienced local extinctions unrelated to the toxic event). These other areas will then not be recolonized, and other local areas that depend on these for colonists will not be recolonized, and so on. Whether landscape-scale propagation of toxic events occurs depends on the landscape structure, the spatiotemporal distribution of the toxic event(s), and the dispersal characteristics and habitat-specific demographics of the organism. In other words, it depends on the interaction between the "natural" spatiotemporal functioning of the population (the extinction-recolonization dynamic) and the spatiotemporal dynamic of the toxic event(s). In general, landscape-scale propagation is more likely to occur the more patchily distributed the population, the more widespread and/or chronic the toxic event, and the more synchronous the occurrence of separate toxic events over the landscape. We propose a method for predicting in particular cases whether landscape-scale propagation of toxic events will occur.

ACKNOWLEDGMENT

We thank Christine Ribic, Tom Lacher, John Emlen and Dean Urban for comments on earlier drafts of this manuscript. This work was supported in part by a Natural Sciences and Engineering Research Council of Canada grant to L. Fahrig and the U.S. Environmental Protection Agency Interagency Agreement (DWCN935524) to Environment Canada for K. Freemark. It has been subjected to the Agency's review and approved for publication.

REFERENCES

Addicott, J.F. 1978. The population dynamics of aphids on fireweed: a comparison of local populations and metapopulations. *Can. J. Zool.* 56: 2554–2564.

Adler, F.R. and B. Nürnberger. 1994. Persistence in patchy irregular landscapes. *Theor. Popul. Biol.* 45: 41–75.

Agur, Z. and J.L. Deneubourg. 1985. The effect of environmental disturbances on the dynamics of marine intertidal populations. *Theor. Popul. Biol.* 27: 75–90.

Ambuel, B. and S.A. Temple. 1983. Area-dependent changes in the bird communities and vegetation of southern Wisconsin forests. *Ecology* 64: 1057–1068.

Barnthouse, L.W. 1992. The role of models in ecological risk assessment: a 1990's perspective. *Environ. Toxicol. Chem.* 11: 1751–1760.

Bennett, A.F. 1987. Conservation of mammals within a fragmented forest environment: the contributions of insular biogeography and autecology. pp. 41–52 in D.A. Saunders, G.W. Arnold, A.A. Burbidge, and A.J.M. Hopkins (eds.). *Nature conservation: the role of remnants of native vegetation.* Surrey Beatty and Sons, Chipping Norton, Australia.

Bennett, A.F. 1990. Habitat corridors and the conservation of small mammals in a fragmented forest environment. *Landsc. Ecol.* 4: 109–122.

Berger, J. 1990. Persistence of different-sized populations: an empirical assessment of rapid extinctions in bighorn sheep. *Conserv Biol.* 4: 91–98.

Best, L.B. 1986. Conservation tillage: Ecological traps for nesting birds? *Wildl. Soc. Bull.* 14: 308–317.

Best, L.B., R.C. Whitmore, and G.M. Booth. 1990. Use of cornfields by birds during the breeding season: The importance of edge habitat. *Am. Midl. Nat.* 123: 84–99.

Blaustein, A.R. 1981. Population fluctuations and extinctions of small rodents in coastal southern California. *Oecologia* 48: 71–78.

Burel, F. 1989. Landscape structure effects on carabid beetles spatial patterns in western France. *Lands. Ecol.* 2: 215–226.

Cairns, J., Jr. 1984. Are single species toxicity tests alone adequate for estimating environmental hazard? *Environ. Monit. Assess.* 4: 259–273.

Dingle, H. 1991. Factors influencing spatial and temporal variation in abundance of the large milkweed bug (Hemiptera: Lygaeidae). *Ann. Entomol. Soc. Am.* 84: 47–51.

Dmowski, K. and M. Kozakiewicz. 1990. Influence of shrub corridor on movements of passerine birds to a lake littoral zone. *Landsc. Ecol.* 4: 99–108.

Duelli, P. 1990. Population movements of arthropods between natural and cultivated areas. *Bio. Conserv.* 54: 193–207.

Dunning, J.B., Jr., B.J. Danielson, and H.R. Pulliam. 1992. Ecological processes that affect populations in complex landscapes. *Oikos* 65: 169–175.

Ehrlich, P.R., R.R. White, M.C. Singer, S.W. McKechnie, and L.E. Gilbert. 1975. Checkerspot butterflies: a historical perspective. *Science* 188: 221–228.

Emlen, J.M. 1989. Terrestrial population models for ecological risk assessment: a state-of-the-art review. *Environ. Toxicol. Chem.* 8: 831–842.

Fahrig, L. 1988. A general model of populations in patchy habitats. *J. Appl. Math. Comput.* 27: 53–66.

Fahrig, L. 1990. Interacting effects of disturbance and dispersal on individual selection and population stability. *Comm. Theor. Biol.* 1: 275–297.

Fahrig, L. 1991. Simulation methods for developing general landscape-level hypotheses of single species dynamics. pp. 417–442 in M.G. Turner and R.H. Gardner (eds.). *Quantitative methods in landscape ecology.* Springer-Verlag, New York.

Fahrig, L. and G. Merriam. 1985. Habitat patch connectivity and population survival. *Ecology* 66: 1762–1768.

Fahrig, L. and G. Merriam. 1994. Conservation of fragmented populations. *Conserv. Biol.* 8: 50–59.

Fahrig, L. and J. Paloheimo. 1987. Interpatch dispersal of the cabbage butterfly. *Can. J. Zool.* 65: 616–622.

Fahrig, L. and J. Paloheimo. 1988a. Determinants of local population size in patchy habitats. *Theor. Popul. Biol.* 34: 194–213.

Fahrig, L. and J. Paloheimo. 1988b. Effect of spatial arrangement of habitat patches on local population size. *Ecology* 69: 468–475.

Fox, G.A., P. Mineau, B. Collins, and P.C. James. 1989. The impact of the insecticide carbofuran (Furadan 480F) on the burrowing owl in Canada. Canadian Wildlife Service (HQ), Technical Report Series No. 72. Environment Canada, Ottawa, ON k1A 0H3, Canada.

Freemark, K.E. 1989. Landscape ecology of forest birds in the Northeast. pp. 7–12 in R.M. DeGraaf and W.M. Healy (compil.). *Is forest fragmentation a management issue in the Northeast?* USDA Forest Service, Northeastern Forest Experiment Station, USA, General Technical Report NE-140.

Freemark, K.E. and C. Boutin, Impacts of agricultural herbicide use on terrestrial wildlife in temperate landscapes: A review with special reference to North America. *Agric., Ecosys. and Environ.*, in press.

Freemark K.E. and B. Collins. 1992. Landscape ecology of birds breeding in termperate forest fragments. pp. 443–454 in J.M. Hagan and D.W. Johnston (eds.). *Ecology and conservation of neotropical migrant landbirds*. Smithsonian Institution Press, Washington, D.C.

Fuller, R., D. Hill, and G.Tucker. 1991. Feeding the birds down on the farm: Perspectives from Britain. *Ambio* 20: 232–237.

Gates, J.E. and L.W. Gysel. 1978. Avian dispersion and fledgling success in field-forest ecotones. *Ecology* 59: 871–883.

Gilpin, M.E. 1990. Extinction of finite metapopulations in correlated environments. pp. 177–186 in B. Shorrocks and I.R. Swingland (eds.). *Living in a patchy environment*. Oxford Science Publications, Oxford, U.K.

Grue, C.E., L.R. DeWeese, P. Mineau, G.A. Swanson, J.R. Foster, P.M. Arnold, J.N. Huckins, P.J. Sheehan, W.K. Marshall, and A.P. Ludden. 1986. Potential impacts of agricultural chemicals on waterfowl and other wildlife inhabiting prairie wetlands: an evaluation of research needs and approaches. *Trans. N.A. Wildl. Nat. Res. Conf.* 51: 357–383.

Hansen, A. and D.L. Urban. 1992. Avian responses to landscape pattern: the role of species life histories. *Landscape Ecol.* 7: 163–180.

Hansen, A., D.L. Urban, and B. Marks. 1992. Avian community dynamics: The interplay of landscape trajectories and species life histories. pp. 170–195 in A.J. Hansen and F. di Castri (eds.). *Landscape boundaries: consequences for biotic diversity and ecological flows. Ecological Studies* 92. Springer-Verlag, NY.

Hanski, I. 1986. Population dynamics of shrews on small islands accord with the equilibrium model. *Biol. J. Linn. Soc.* 28: 23–36.

Hanski, I. 1991. Single-species metapopulation dynamics: concepts, models and observations. *Biol. J. Linn. Soc.* 42: 17–38.

Harrison, S. and J.F. Quinn. 1989. Correlated environments and the persistence of metapopulations. *Oikos* 56: 293–298.

Henein, K. and G. Merriam. 1990. The elements of connectivity where corridor quality is variable. *Landsc. Ecol.* 4: 147–170.

Herben, T., H. Rydin, and L. Söderström. 1991. Spore establishment probability and the persistence of the fugitive invading moss, *Orthodontium lineare*: a spatial simulation model. *Oikos* 60: 215–221.

Howe, R.W., G.J. Davis, and V. Mosca. 1991. The demographic significance of sink populations. *Bio. Conserv.* 57: 239–255.

Jorgensen, S.W. 1990. *Modelling in ecotoxicology. Developments in Environmental Modelling 16*. Elsevier, New York.

Kareiva, P. 1990. Population dynamics in spatially complex environments: theory and data. *Philos. Trans. R. Soc. London B 330:* 175–190.

Lande, R. 1988. Demographic models of the northern spotted owl (*Strix occidentalis caurina*) *Oecologia* 75: 601–607.

Lawton, J.H. and G.L. Woodroffe. 1991. Habitat and the distribution of water voles: why are there gaps in a species' range? *J. Anim. Ecol.* 60: 79–91.

Lefkovitch, L.P. and L. Fahrig. 1985. Spatial characteristics of habitat patches and population survival. *Ecol. Modelling* 30: 297–308.

Levins, R. 1969. Some demographic and genetic consequences of environmental heterogeneity for biological control. *Bull. Entomol. Soc. Am.* 15: 237–240.

Lidicker, W.Z., Jr. 1975. The role of dispersal in the demography of small mammals. pp. 103–128 in F.B. Golley, K. Petrusewicz. and L. Ryskowski (eds.). *Small mammals: their productivity and population dynamics.* Cambridge University Press: New York.

Lynch, J.E. and D.F. Whigham. 1984. Effects of forest fragmentation on breeding bird communities in Maryland, USA. *Biol. Conserv.* 28: 287–324.

Mader, H.-J., C. Schell, and P. Kornacker. 1990. Linear barriers to arthropod movements in the landscape. *Biol. Conserv.* 54: 209–222.

Mason, L.G. 1977. Extinction, reproduction and population size in natural populations of ambushbugs. *Evolution* 31: 445–447.

McMullen, L.H. and M.D. Atkins. 1962. On the flight and host selection of the Douglas-fir beetle, *Dendroctonus pseudotsugae* Hopk. (Coleoptera: Scolytidae). *Can. Entomol.* 94: 1309–1325.

Menges, E.S. 1990. Population viability analysis for an endangered plant. *Conserv. Biol.* 4: 52–62.

Merriam, G. and J. Wegner. 1992. Local extinctions, habitat fragmentation and ecotones. pp. 150–169 in A.J. Hansen and F. Di Castri (eds.). *Landscape boundaries.* Springer-Verlag, NY.

Noon, B.R. and C.M. Biles. 1990. Mathematical demography of spotted owls in the Pacific Northwest. *J. Wild. Manage.* 54: 18–27.

O'Connor, R.J. and M. Shrubb. 1986. *Farming and birds.* Cambridge University Press, London, U.K.

Paine, R.T. 1988. Habitat suitability and local population persistence of the sea palm *Postelsia palmaeformis. Ecology* 69: 1787–1794.

Potter, M.A. 1990. Movement of North Island brown kiwi (*Apteryx australis mantelli*) between forest fragments. *N. Z. J. Ecol.* 14: 17–24.

Prokopy, R.J., R.H. Collier, and S. Finch. 1983. Visual detection of host plants by cabbage root flies. *Entomol. Exp. Appl.* 34: 85–89.

Pulliam, H.R. 1988. Sources, sinks, and population regulation. 1988. *Am. Nat.* 132: 652–661.

Pulliam, H.R. 1994. Incorporating concepts from population and behavioral ecology into models of exposure to toxins and risk assessment. pp. 13–26 in R.J. Kendall and T.E. Lacher (eds.). *Wildlife toxicology and population modeling: integrated studies of agroecosystems.* Lewis Publishers, Boca Raton, FL.

Reddingius, J. and P.J. den Boer. 1970. Simulation experiments illustrating stabilization of animal numbers by spreading of risk. *Oecologia* 5: 240–284.

Roff, D.A. 1974a. Spatial heterogeneity and the persistence of populations. *Oecologia* 15: 245–258.

Roff, D.A. 1974b. The analysis of a population model demonstrating the importance of dispersal in a heterogeneous environment. *Oecologia* 15: 259–275.

Rogers, C.A. and K.E. Freemark. 1991. *A feasibility study comparing birds from organic and conventional (chemical) farms in Canada.* Canadian Wildlife Service (HQ), Techincal Report Series No. 137. Environment Canada, Ottawa, ON K1A 0H3, Canada.

Seno, H. 1988. Effect of a singular patch on population persistence in a multi-patch system. *Ecol. Modelling* 43: 271–286.

Sheehan, P.J., A. Baril, P. Mineau, D.K. Smith, A. Harfenist, and W.K. Marshall. 1987. *The impact of pesticides on the ecology of prairie nesting ducks.* Canadian Wildlife Service (HQ), Techincal Report Series No. 19. Environment Canada, Ottawa, ON K1A 0H3, Canada.

Simberloff, D., J.A. Farr, J. Cox, and D.W. Mehlman. 1992. Movement corridors: conservation bargains or poor investments? *Biol Conserv.* 6: 493–504.

Sotherton, N.W. 1991. Conservation headlands: A practical combination of intensive cereal farming and conservation. pp. 373–397 in L.G. Firbank, N.Carter, J.F. Darbyshire, and G.R. Potts (eds.). *The ecology of temperate cereal fields.* Blackwell Scientific Publications, London, U.K.

Streibing, J.C. 1988. Weeds — the pioneer flora of arable land. *Ecol. Bull.* (Copenhagen) 39: 59–62.

Temple, S.A. and J.R. Cary. 1988. Modeling dynamics of habitat-interior bird populations in fragmented landscapes. *Conserv. Biol.* 2: 340–347.

Tew, T.E., D.W. MacDonald, and M.R.W. Rands. 1992. Herbicide application affects microhabitat use by arable wood mice (*Apodemus sylvaticus*). *J. Appl. Ecol.* 29: 532–539.

Thomas, C.D. 1991. Spatial and temporal variability in a butterfly population. *Oecologia* 87: 577–580.

Urban, D.L. and H.H. Shugart, Jr. 1986. Avian demography in mosaic landscapes: Modeling paradigm and preliminary results. pp. 273–279 in J. Verner, M.L. Morrison, and C.J. Ralph (eds.). *Wildlife 2000: Modeling habitat relationships of terrestrial vertebrates.* University of Wisconsin Press, Madison, WI.

Vance, R.R. 1980. The effect of dispersal on population size in a temporally varying environment. *Theor. Popul. Biol.* 18: 343–362.

Vance, R.R. 1984. The effect of dispersal on population stability in one-species discrete-space population growth models. *Am. Nat.* 123: 230–254.

Van der Meijden, E., T.J. de Jong, P.G.L. Klinkhamer, and R.E. Kooi. 1985. Temporal and spatial dynamics of biennial plants. pp. 91–103 in J. Haeck and J.W. Woldendorp (eds.). *Structure and functioning of plant populations 2.* North-Holland, Amsterdam.

Verboom, B. and R. van Apeldoorn. 1990. Effects of habitat fragmentation of the red squirrel, *Sciurus vulgaris* L. *Landsc. Ecol.* 4: 171–176.

Verboom, J., A. Schotman, P. Opdam, and J.A.J. Metz. 1991. European nuthatch metapopulations in a fragmented agricultural landscape. *Oikos* 61: 149–156.

Villard, M.-A., K.E. Freemark, and G. Merriam. 1992. Metapopulation dynamics as a conceptual model for neotropical migrant birds: an empirical investigation. pp. 474–482 in J.M. Hagan and D.W. Johnston (eds.). *Ecology and conservation of neotropical migrant landbirds.* Smithsonian Institution Press, Washington, D.C.

Wallace, G.K. 1958. Some experiments on form perception in the nymphs of the desert locust, *Schisterocerca gregaria* Forskal. *J. Exp. Biol.* 35: 765–775.

CHAPTER 12

Progress in Toxicity Testing — An Academic's Viewpoint

Kenneth L. Dickson

INTRODUCTION

In the over 20 years that I have worked in aquatic toxicology, I have seen it become a recognized field of environmental sciences. The reason for its acceptance is that it has offered practical methods, approaches, and data useful for making decisions about the ecological effects of chemicals and complex effluents. This is not to say that aquatic toxicology provides definitive data or that there cannot be improvement. Certainly most will agree that the field of ecotoxicology is still in early to mid stages of development and that the field is dynamic. Ecotoxicology had its roots in organismal level testing and has evolved both up and down the scale of biological organization. Recent developments in the field have been made in the areas of biochemical and physiological biomarkers, as well as in multispecies testing (micro-, meso-, and multicosms). It is my premise that ecotoxicology methods and approaches are needed at all levels of biological organization because they all provide useful information needed to make prudent decisions about the effects of chemicals on organisms, populations, communities, and ecosystems. The challenge facing those working in the field of ecotoxicology is to settle on a group of ecologically important assessment endpoints (particularly in multispecies tests) that can be reliably measured and, thus, can be used in ecological risk assessments.

When I first became aware of the use of aquatic toxicity testing in the mid 1960s, it had only recently left the era of "pickle jar" fish testing pioneered by Hart, Doudoroff and Greenbank.[1] Chemicals and effluent were evaluated for acute toxicity, and lethality was the dominant endpoint of choice. The assessment

0-87371-599-3/95/$0.00+$.50
© 1995 by CRC Press, Inc.

endpoint (lethality) was also the measurement endpoint of the test (see Chapter 3). It was assumed that the species used in toxicity testing were surrogates for sensitive species exposed in nature and that if concentrations of chemicals and/ or effluent found not to cause lethality in the laboratory were not exceeded in nature that acute lethality would not occur. Single species acute toxicity tests became the foundation for aquatic toxicology. Why? The tests were simple to conduct, relatively inexpensive, and easily interpreted. However, many appropriately began to question whether or not they provided adequate information about the effects of chemicals to be truly protective of aquatic life.

Concerns over long-term (chronic) effects of exposures to low levels of chemicals on the growth and reproduction of organisms stimulated a number of investigators to develop single species life cycle methods and approaches addressing these concerns.[2-4] Methods were developed to assess the chronic effects of chemicals and effluent on both freshwater and marine organisms and terrestrial fauna. When applied to chemicals and effluent, these methods allowed derivation of No Observable Effect Levels (NOELs) using growth and reproduction as the endpoints. In the aquatic toxicity area, full life cycle single species testing further evolved to short term chronic assays because it was found that early stages of the life cycle were frequently the most sensitive to many stressors and because of time and expense of conducting full life cycle chronic toxicity tests.[5,6] Both the full life cycle tests and short-term chronic tests provided the aquatic toxicologist with important tools to screen chemicals and effluent for potential effects on reproduction and growth. These tests were incorporated into state and federal regulatory programs to assess the potential ecological effects of pesticides (Federal Insecticide, Fungicide and Rodenticide Act, FIFRA), the toxicity of chemicals used in commerce (Toxics Substance Control Act, TSCA), and the toxicity of industrial and muncipal effluents (Clean Water Act). These toxicity tests have unambiguous assessment and measurement endpoints, are relatively simple to conduct, and produce reliable results.[7]

Concomitant with developments in single species and chronic toxicity testing was the development of methods and approaches to assess the effects of stressors (toxicants, nutrients, heat, etc.) on assemblages of organisms that more accurately reflect the "real world". These activities were stimulated by a variety of concerns about how protective were decisions based on single species toxicity test results (i.e., were laboratory test species good surrogates?) and interest in understanding how stressors affect structure and function of ecosystems. A variety of multispecies toxicity testing approaches have been developed ranging from microbial tests[8,9] to ecosystem level mesocosm tests.[10-13] Cairns[14] summarizes many of the developments in multispecies testing and discusses strengths and weaknesses. One of the apparent challenges facing multispecies testing is developing a consensus on which endpoints are to be used.

Spawned by the passage of TSCA in the U.S. (TSCA, 1972), the hazard assessment/risk assessment paradigm began its development in the early to mid

1970s. TSCA required that new chemicals and/or new uses of existing chemicals be evaluated for potential effects on humans and environmental health by the U.S. Environmental Protection Agency (U.S. EPA) *prior* to their use in commerce. Unlike FIFRA, which has a minimum data set requirement, TSCA had none. This stimulated U.S. EPA, individuals in academia, industry, and environmental groups to devise approaches for doing hazard assessments. What evolved from a number of workshops, conferences, and technical papers was the hazard assessment paradigm.[15–17] The essence of this paradigm is a comparison of the Estimated Environmental Concentration (EEC) with an estimate of the Effects Concentration (EC). The paradigm brings together both fate and effects information on a chemical to arrive at a decision about the probable environmental consequences of its use. As the hazard assessment paradigm developed, a tiered testing protocol emerged where simple methods were used in early stages of assessment to estimate the EEC and EC with more complex and expensive methods used in later stages, if needed. Decision criteria based on the quotient method (EC/EEC) were developed to guide progression through the assessment. If the EEC was very near the EC then higher tier (more complex) testing was indicated. However, if the EEC was much lower than the EC, then a decision about the hazard (safety) of the chemical could be made without proceeding to higher tiers of testing. This hazard assessment paradigm has been the foundation of U.S. EPA's approach for requiring ecotoxicity testing of pesticides and other chemicals.

The quotient approach depends on professional judgment to make decisions about the hazard of a chemical and/or to require additional higher tier testing. This limitation has stimulated several investigators to explore methods of quantifying the uncertainty associated with the EEC and the EC.[18–20] The goal of these endeavors has been to develop the tools to do quantitative ecological risk assessment. The difference between hazard assessment and ecological risk assessment is that ecological risk assessment has as its goal to place a probability on the occurence of an ecological effect if a concentration of a chemical in the environment is exceeded. The foundations of the emerging methods of ecological risk assessment are results of single species toxicity testing and population models. It is unclear how results of multispecies toxicity testing will be used in these endeavors. However, multispecies tests undoubtedly will be important in validation of ecological risk asessment models.

In summary, a great deal of progress has been made in toxicity testing in the last 20 years. We now have toxicity testing methods to assess chronic toxicity, effects on populations, and ecosystem structure and function. Biomarkers of exposure and effects have been developed that hold promise of providing not only a better mechanistic understanding of how toxicants affect organisms, but also may serve as early warnings of effects and, in some cases, provide diagnosis of the causes of stress. What is lacking, in my opinion, is the development of a consensus among ecotoxicologists, ecologists, and risk assessors about what assessment endpoints will be used in the future to make decisions about the health of the environment. It is this point that I want to address in the remainder of this short discussion.

PERSPECTIVE

Approaches and methods developed for toxicity testing have to be cognizant of what is going on in the fields of ecological risk assessment and environmental monitoring. The historical goal of ecotoxicology has been to predict the effects of stressors on different levels of biological organization. There is a need for methods and approaches at all levels of biological organization because results of tests at each level of organization provide different but useful information in assessing the potential and real risks of chemicals. For example, biochemical biomarkers, such as mixed function oxidases, stress proteins, DNA adducts, acetylcholinesterase and immunosuppression, etc., which operate at the subcellular level of biological organization, have been shown to be sensitive indicators of exposure and effects.[21] They can be useful in mapping out contaminated areas at hazardous waste sites, as well as assessing sublethal impacts on exposed wildlife. Some, such as metallothionein, ALAD, and stress proteins appear to have value in the diagnosis and identity of what is causing the stress.[21-23]

Approaches and methods at the organismal level of ecological organization have historically been the foundation of ecotoxicology. In my opinion, it is unlikely that this will change. People are concerned about the survival and population success of organisms. The endpoints are relatively easy to communicate to decisionmakers and the public. People can identify with survival, growth, and reproduction, all typical endpoints of ecotoxicity testing at the organismal level. By comparison, it is hard for decisionmakers and the public to identify with or understand the significance of an enhanced mixed function oxidase activity caused by exposure to neutral organic compounds or a depression of the rate of carbon cycling in a mesocosm caused by a pesticides impact on zooplankton. Other reasons supporting my feeling that organismal level testing will continue to be the foundation of ecotoxicolgy relate to both my pragmatic nature and recent experiences examining the results of a series of field studies conducted by the U.S. EPA and others. From a pragmatic standpoint, the value of the information derived compared to the cost of conducting the tests makes organismal level testing attractive. Single species toxicity tests are significantly less expensive than microcosm and mesocosm multispecies tests and provide easily interpretable results. The biggest draw back on organismal level testing may be the increasing ethical concerns over the use of animals and plants in testing. These concerns also apply to multispecies testing. In my opinion, future directions in organismal level testing should focus on ways to minimize the use of animals and plants and on further developing population, community, and ecosystem models, which use results of single species toxicity tests as inputs.

Recently, my colleagues and I completed a study for the U.S. EPA, the purpose of which was to examine the relationship(s) between ambient toxicity and instream community responses.[24] In the mid 1980s, the U.S. EPA conducted a series of field studies in aquatic systems receiving effluent from municipalities and industries.[25] In these studies, field assessments of various aquatic communities (fish, benthic macroinvertebrates, zooplankton, phytoplankton, etc.) were conducted and ambient toxicity assessed using *Ceriodaphnia dubia* 7-day chronic toxicity test and the fathead minnow 7-day larval growth toxicity test. A study that we conducted on the Trinity River in Texas used field studies to assess impacts of municipal discharges on aquatic life and evaluated ambient toxicity to the same test species. The results of analyzing these combined data showed that there was a relationship between ambient toxicity as predicted by organismal level toxicity tests and instream community response. Interestingly, fish species richness showed the strongest relationship. While it is easy to envision sites or situations where it would be difficult to show such a relationship empirically, for example, where the toxicity from the effluent is low and/or where habitat is limiting, the result of the study significantly increased my confidence that these organismal level toxicity tests are predictive and protective of effects in the "real world".

This brings me to the question of what are the needs for multispecies toxicity tests. As mentioned earlier it is my premise that we need toxicity testing approaches at all levels of biological organization. This includes multispecies approaches such as microcosms, mesocosms, field studies, and whole ecosystem investigations. However, I feel that their primary role is in the validation of predictions made from toxicity testing conducted at lower levels of biological organization and in validation of ecological models that extrapolate results from single species toxicity testing to population, community, and ecosystem levels. In addition, I feel that multispecies toxicity tests such as microcosms and mesocosms can potentially make a significant contribution to the hazard assessment of pesticides and other chemicals whose estimated environmental exposure concentration approximates known effects levels. Use of multispecies level tests are particularly valuable for high volume, high benefit chemicals that also have high predicted toxicity based on single species toxicity test results. There are several barriers, however, that must be overcome before this potential can be realized. The experimental design of microcosm and mesocosm experiments needs to be more carefully considered. Many of the mesocosm toxicity tests that have been conducted in support of pesticide registration have used an experimental design that limits the final interpretation of the data (see Chapter 9). Perhaps most importantly, we need to develop a consensus of the effects indicators that are going to be monitored in the mesocosm experiments around which hazard/risk assessment decisions are going to be based. Historically, mesocosms have been required by the U.S. EPA to support pesticide (particularly pyrethoids) registration and reregistration when the environmental exposure concentration approximates the environmental effects concentration. However, the U.S. EPA has recently softened its requirement for mesocosm testing because the U.S. EPA decided that data developed are difficult to interpret and add little value to the hazard/risk assessment.[26] It is unclear how the results of the mesocosm toxicity tests were evaluated by the U.S. EPA. It would be unfortunate if the appropriate use of microcosms and mesocosms was abandoned in pesticide risk assessment because of flawed designs and/or analyses. Thus, it is essential that the scientific basis used to determine that mesocosm data add little to the risk assessment be made public so that this decision can be evaluated by all those interested in multispecies testing and evaluation.

As Cairns[27] has pointed out on a number of occasions, ecosystem management requires both a predictive (toxicity testing) and a monitoring component providing feedback. Currently, there are several initiatives underway globally to monitor status and trends in ecological resources. While none of these efforts provide real-time feedback, they do have the potential to provide information about the effectiveness of environmental policy and regulation. The U.S. EPA-directed Environmental Monitoring and Assessment Program (EMAP) is one of the programs with which ecotoxicologist should become very familiar, since it is logical that there be a relationship between effects indicators used to monitor status and trends of ecological resources and effects indicators used by ecotoxicologists in their toxicity tests. I think it is particularly important that those ecotoxicologists who are conducting multispecies toxicity tests carefully examine the assessment endpoints that they are using and attempt to relate them to those used by monitoring programs.

CONCLUSIONS

Ecotoxicology, as a branch of environmental science, has developed at a rapid rate. There now exist methods and approaches for evaluating the effects of stressors at all levels of biological organization. These assessment tools can all contribute to our understanding of the effects of chemicals and other stressors. Similarly, they contribute to the development of knowledge about homeostatic mechanisms in nature. Rather than argue about the relative value of toxicity testing at one level of biological organization vs. another level, it would be far more constructive to focus on understanding their interrelationships. Out of this quest will come a better understanding of how nature works, as well as better methods and approaches for predicting and preventing ecological harm from anthropogenic stressors.

It is my belief that ecotoxicologists have an important contribution to add to the debate about what facets of the biotic environment are important to protect. It is becoming increasingly clear that hard choices will have to be made about ecological indicators of ecosystem health. I encourage my fellow ecotoxicologists, particularly those working at the multispecies level of organization, to become involved in this debate, to share their wisdom and experience and to evaluate the efficacy of proposed ecosystem indicators of environmental health.

REFERENCES

1. Hart, W. B., P. Doudoroff, and J. Greenbank, *The Evaluation of the Toxicity of Industrial Wastes, Chemicals and other Substances to Fresh-water Fishes, Waste Control Laboratory,* Philadelphia, PA: Atlantic Refining Company (1945).
2. Biesinger, K. E., *Recommended Bioassay Procedure for Daphnia magna Chronic Tests in a Standing System,* Tentative Procedure, Duluth MN: U.S. Environmental Protection Agency (1973).
3. Sprague, J. B., The ABC's of Pollutant Bioassay Using Fish, in *Biological Methods for the Assessment of Water Quality,* STP528, J. Cairns, Jr., and K. L. Dickson, Eds., Philadelphia, PA: American Society for Testing and Materials (1973), pp. 6–30.
4. Schimmel S. C. and D. J. Hansen, Sheepshead Minnow (*Cyprinodon variegatus*): An Estuarine Fish Suitable for Chronic (Entire Life-Cycle) Bioassay, in *Proceedings of the 28th Annual Conference,* Southeast Game Fish Commission (1974), pp. 392–398.
5. McKim J. M., Evaluation of Test With Early Life Stages of Fish for Predicting Long-term Toxicity, *J. Fish. Res. Board Can.* 34:1148–1154 (1977).
6. Horning, W. B. and C. I. Weber, *Short-term Methods for Estimating the Chronic Toxicity of Effluent and Receiving Waters to Freshwater Organisms,* EPA/600/4–85/014, Cincinnati, OH: U.S. EPA Office of Research and Development, Environmental Monitoring and Support Laboratory (1985).
7. U.S. Environmental Protection Agency, *Technical Support Document for Water Quality-based Toxics Control,* EPA 505/2–91/001, Washington, DC: Office of Water (1985).
8. Taub, F. B., Demonstration of Pollution Effects in Aquatic Microcosm, *Int. J. Environ. Stud.* 10:23–33 (1976).
9. Cairns, J., J. R. Pratt, B. R. Niederlehner, and P. V. McCormick, A Simple, Cost-effective Multispecies Toxicity Test Using Organisms With A Cosmopolitan Distribution, *Environ. Monit. Assess.* 6:207–220 (1986).
10. Howick, G. L., F. deNoyelles, J.M Giddings, and R. L. Graney, Earthen Ponds Versus Fiberglass Tanks as Venues for Assessing the Impact of Pesticides on Aquatic Environments: A Parallel Study with Sulprofos, in *Aquatic Mesocosm Studies in Ecological Risk Assessment,* R. L Graney, J. H. Kennedy, and J. H Rodgers, Jr., Eds., Boca Raton, FL: Lewis Publishers (1993).
11. Crossland, N. O. and T. W. La Point, The Design of Mesocosm Experiments, *Environ. Toxicol. Chem.* 11(3) 1–4 (1992).
12. Genter, R. B., D. S. Cherry, E. P. Smith, and J. Cairns, Jr., Algal-Periphyton Population and Community Changes From Zinc Stress in Stream Mesocosms, *Hydrobiologia* 153:261–175 (1987).
13. Touart, L. W., *Aquatic Mesocosm Testing To Support Pesticide Registration,* Hazard Evaluation Division Technical Guidance Document, EPA-540/09–88–035, Washington, DC: U.S. Environmental Protection Agency (1987).

14. Cairns, J., Jr., *Multispecies Toxicity Testing,* Boca Raton, FL: Lewis Publishers (1985).
15. Cairns, J., Jr., K. L. Dickson, and A. W. Maki, Eds., *Estimating The Hazard of Chemical Substance to Aquatic Life*, STP657, Philadelphia, PA: American Society for Testing and Materials (1978).
16. Dickson, K. L., A. W. Maki, and J. Cairns, Jr., Eds., *Analyzing the Hazard Evaluation Process,* Washington DC: American Fisheries Society (1979).
17. Bergman, H. L., R. A. Kimerle, and A. W. Maki, Eds., *Environmental Hazard Assessment of Effluent,* New York: Pergamon Press (1986).
18. Barnthouse, L. W. and G. W. Suter, II, Eds., *User's Manual for Ecological Risk Assessment*, ORNL-6251, Oak Ridge, TN: Oak Ridge National Laboratory (1986).
19. Suter, G. W., II, *Ecological Risk Assessment,* Boca Raton, FL: Lewis Publishers (1992).
20. Bartell, R., H. Gardner, and R. V. O'Neill, *Ecological Risk Estimation,* Boca Raton, FL: Lewis Publisher (1992).
21. Peakall, D., *Animal Biomarkers as Pollution Indicators,* New York: Chapman & Hall (1992).
22. Sanders, B. M., Stress Proteins: Potential as Multitiered Biomarkers, in *Biological Markers of Environmental Contamination*, J. McCarthy and L. Shugart, Eds., Boca Raton, FL: CRC Press (1990), pp. 135–161.
23. Dyer, S. D., K. L. Dickson, E. G. Zimmerman, and B. M. Sanders, Tissue Specific Patterns of Heat Shock Protein Synthesis and Thermal Tolerance of the Fathead Minnow (*Pimephales promelas*), *Can. J. Zool.* 69:2021–2027 (1991).
24. Dickson, K. L., W. T. Waller, J. H. Kennedy, and L. P. Ammann, Assessing the Relationship between Ambient Toxicity and Instream Biological Response, *Environ. Toxicol. Chem.* 11:1207–1322 (1992).
25. Mount, D. I., N. A. Thomas, T. J. Norberg, M. T. Barbour, T. H. Roush, and W. F. Brandes, *Effluent and Ambient Toxicity Testing and Instream Community Response on the Ottawa River, Lima, Ohio*, EPA 600/3–84–080, Duluth, MN: U.S. Environmental Protection Agency (1984).
26. U. S. Environmental Protection Agency, Program Guidance on Ecological Risk Management, Washington, DC: Office of Pesticides and Toxic Substances (1992).
27. Cairns, J., Jr., and T. V. Crawford, Eds., *Integrated Environmental Management,* Boca Raton, FL: Lewis Publishers (1991).

CHAPTER 13

Future Trends in Ecotoxicology

John Cairns, Jr.

Two events — the continuing development of a global marketplace and our concomitant entry into an "information age" — are already in place that will continue to affect the entire field of environmental management, including toxicity testing at different levels of biological organization. These are inextricably linked phenomena, but each will have somewhat different effects on environmental management.

THE IMPACT OF ENTERING THE INFORMATION AGE

Toffler[1,2] has described three waves of cultural change: the agricultural revolution, the industrialization revolution, and a postindustrial "information age," in which knowledge-based technologies (those dependent on gathering, integrating, and disseminating information) replace smokestack industries as the primary source of capital creation and power. Maynard and Mehrtens[3] extend this concept by describing a fourth cultural wave of change in which the business community embraces global stewardship.

The computer age has brought an accessibility to specialty information that is previously unprecedented. At present, information gathered by toxicity tests is only accessible to the initiated. Such limitations are faults common to almost every discipline or area of specialization. However, information not easily integrated with other dissimilar types of information will be far less influential in policymaking than information that is. Consequently, the ability to integrate a specific type of information with a vast array of information of a dissimilar nature is a basic requirement of the information age. During the development of the field of environmental toxicology, the primary forcing factor has been compliance with regulatory statutes. Thus, while not without ambiguities, the

0-87371-599-3/95/$0.00+$.50

type of information and the style of presentation have been fairly straightforward and consistent. However, the form ideal for regulation and regulators may not be easily incorporated as one of multiple factors in policy and decisionmaking. All information generators wishing to be effective in the information age will face similar problems of incongruence. Instead of compliance (which deals with essentially a yes or no answer), communication should deal with a graded series of risks, ecological breakpoints, or thresholds. These must be considered in conjunction with a variety of other risks that are no less important but based on information generated in a different way. The nature of the problem to be solved, the way in which the information will be used, and the level of toxicological literacy of the persons required to integrate the information are crucial to the overall success of our efforts. Rather than leave these considerations to others, ecotoxicologists will undoubtedly have a greater impact and more influence on policy and decisionmaking if these factors are considered when the information is provided to larger databases. This does not mean altering the information in an unscientific way but, rather, modifying it so that it is compatible with other types of information. This is the "stuff" of ecological risk assessment[4,5] and risk management.

Some situations seem abundantly clear about the effects of the information age on ecotoxicology.

1. Information will require site-specific relevance — this means that standardized tests must either be modified into highly site-specific tests using indigenous organisms, or the information obtained must be extrapolated across conditions and scale to provide information relevant to site-specific decisionmaking. Both approaches require considerable professional judgment.
2. Information will have to be generated rapidly in order to meet deadlines other than the simple, highly scheduled deadlines of regulatory agencies.
3. Feedback loops will be far more complex when additional information is required, most likely with specific instructions on the nature of this information, the level of detail or precision required, and the time period within which the information will be useful and beyond which it may be totally inappropriate or useless.

THE EFFECTS OF GLOBALIZATION ON ECOTOXICOLOGY

Society has often failed in the past to grasp the concept of the global community without suffering serious consequences. However, current attempts to think locally and act locally only (in contrast to Rene Dubos' famous prescription to "think globally, act locally") will have serious adverse consequences, just as ignoring unique characteristics and attributes of local systems will. This is an age of interconnectedness, both economically and ecologically. Even though quite distant from the Chernobyl nuclear power plant explosion, 100,000 reindeer in Lapland were slaughtered because of radioisotopes they acquired. One contaminated air plume traveled from Chernobyl to the area inhabited by the reindeer (Bruce Wallace, personal communication and Levi[6]). Similarly, fruits, vegetables, and other items of human diet were contaminated in the Balkans by another plume from the same source. Song birds in the U.S. are suffering diminishing numbers, possibly because wintering ground habitats south of the U.S. border are disappearing at an even greater rate than habitats in the U.S. Preservation of these species will require, if not a global outlook, at least a hemispheric outlook. Many other species, particularly whales, require a global outlook for survival. Maintenance of the gas balance in the thin envelope surrounding the earth will also require a greatly broadened outlook because circulation of contaminants introduced on one continent frequently, perhaps inevitably, reaches other continents.

The field of ecotoxicology is not yet well defined.[7] However, substantial demands are now being made on practitioners and will increase in intensity. A major use of ecotoxicology and other types of environmental information in a global context will probably be for political negotiations. This is quite a different role for at least that part of toxicology that is compliance driven, which is arguably the largest single component at present. Perhaps, the negotiation will be based on a more sophisticated formulation of the I=PAT equation.[8,9] In this equation, which is based on energy consumption, environmental impact (I) = population size X level of affluence X technology. A comparable equation could probably be developed using per capita toxic waste generation released into the environment or some other measure of environmental stress. At present, I believe that energy use is a common denominator for all components of human society because, even in undeveloped countries, energy is at the center of life. Furthermore, society has a higher energy literacy level than its ecotoxicological literacy level, and, of course, environmental literacy as a whole is arguably somewhere between the two. Without question, no society has adequate literacy in these important areas for the sorts of judgments that will be necessary for the globalization of human society, let alone making the proper decisions by nationality. Elected representatives or others in power seldom have any better knowledge in any of these areas than the average citizen. The U.S., for example, with its comparative affluence has a high per capita energy consumption and a considerable output of hazardous substances. Its environmental impact, due to level of affluence and types of technology (including agricultural), is much higher per capita than the environmental impact per capita of the People's Republic of China, even though it has a quarter of the world's population. However, the environmental impact of both nations may be quite similar due to the total population numbers in China and the high per capita impact in the U.S. The accuracy of this concept cannot be tested because our means of making quantitative measurements are still quite rudimentary. However, a simplistic approach for illustrative purposes is to limit global pollution on the basis of geographic area — each country could then determine the appropriate balance of the components of the I=PAT equation to limit its global pollution to a predetermined level. Forcing such an agreement may be virtually impossible since, as B. F. Skinner[10] has noted, behavior is unlikely to change unless people have experienced some consequences. Since the United Nations is a weak agency, the consequences will probably take the form of a catastrophe caused by overutilization of resources or excessive production of toxicological materials, or both.

CONCLUSIONS

As always, unexpected events can have dramatic short-term and possibly even long-term impacts. When the Russians launched Sputnik, the first orbiting spacecraft, it so shocked policymakers in the U.S. (that Russia could be ahead of it in a scientific/engineering endeavor) that substantial amounts of money were put into not only the space program but also science/engineering education as a whole, particularly at the university and college levels. A major environmental disaster might produce similar results. Nevertheless, trends in science may well be overshadowed by a changing *ethos* (a set of guiding beliefs) in human society toward natural systems. This seems almost unthinkable in a year when the spotted owl controversy in the logging areas of the American Pacific Northwest generated bumper stickers such as "Kill a spotted owl — Save a logger's job" and worse. Ecotoxicological information can be used to protect ecosystems and, to a degree, ecological and related evidence can be used to improve ecosystem condition. However, empathy is required to make them flourish. For robust ecosystem health, human society must organize its activities voluntarily so that truly wild systems can endure, and all ecosystems, or at least the majority, can flourish. To accomplish this, environmental literacy throughout the society must be high, and there must be a sensitivity to the well being of natural systems that is more instinctive than scientific, although this introduces an area in which responses to observations are difficult to quantify or verbalize. Many years ago, in what is now the U.K., there were river wardens who walked along the rivers and judged their condition. Since each was assigned to a particular stretch of river as essentially a lifetime assignment, each gained a high degree of familiarity with the ecological health of the respective river and sensitive to unfavorable change. These wardens were not scientists in the sense of having formal degrees, but they were apparently astute practicing ecologists nevertheless. However, in an era when nearly half the global human population is urbanized (and the halfway mark is expected to be exceeded before the year 2000), familiarity with natural systems is understandably markedly diminished, and, thus, environmental literacy of the type acquired by the river wardens is rarer. Perhaps a partnership between human society and nature will be established — the needs of other species and the complex multivariate systems that they require will become a part of the *ethos* of human society. If this happens in a substantive way, compliance monitoring will still occur but will involve a different set of endpoints and "evidence" than the type ecotoxicologists now consider routine. Only a decade or so ago, such a possibility would have been considered remote by most scientists and, by many, visionary. However, some present trends indicate that human population growth is no longer as inevitable as it once appeared, and, while the problems of hazardous waste sites are intimidating, the development of green engineering suggests that such sites will no longer be appearing at their past rates.

These are exciting changes indeed and confirm the now fairly well-established belief that education will not stop upon graduation with whatever the highest degree happens to be but continue throughout an individual's lifetime. The connections between academe, industry, and government, once tenuous in many ways, will strengthen as individuals move among the three units with far more frequency and time spent than before. The possibilities of cooperation rather than polarization thus seem enhanced and, if all three components of the triad embrace continual professional renewal and the inevitability of rapid change, the entrenched positions characteristic of the last 4 decades seem unlikely to persist in their present forms. In short, the position of "them" and "us" so characteristic of the past will probably diminish, perhaps disappear entirely. At the first Earth Day, the "them" was the industrial component of society that produced the goods upon which human society depends, and the "us" was the people not directly involved in producing the goods — some in academe, some independently wealthy, and some in service industries that characteristically produce few toxic waste materials. Another "them" and "us" is the perceived dichotomy between humans and all other species. Those who support natural systems ("them") are automatically assumed to be against "us" (human society). In fact, all species are "us", even though the present level of environmental literacy may not be sufficiently high to acknowledge this. Unfortunately, catastrophic events may make this clearer in the future or, more hopefully, environmental literacy will improve sufficiently to make catastrophic events unnecessary for a major paradigm shift to occur in human society's relationship with natural systems.

REFERENCES

1. Toffler, A., *The Third Wave,* New York: Bantam Books (1988).
2. Toffler, A., *Powershift,* New York: Bantam Books (1990).
3. Maynard, H. B., Jr. and S. E. Mehrtens, *The Fourth Wave: Business in the 21st Century,* San Francisco, CA: Berrett-Koehler Publishing (1993).
4. National Research Council, *Risk Assessment in the Federal Government: Managing the Process,* Washington, DC: National Academy Press (1983).
5. Norton, S. B., D. J. Rodier, J. H. Gentile, W. H. van der Schalie, W. P. Wood, and M. W. Slimak, A Framework for Ecological Risk Assessment at the EPA, *Environ. Toxicol. Chem.* 11:1663–1672 (1992).

6. Levi, H. W., Radioactive Deposition in Europe After the Chernobyl Accident and Its Long-term Consequences, *Ecol. Res.* 6:201–216 (1991).
7. Cairns, J., Jr. "Will the Real Ecotoxicologist Please Stand Up?" *Environ. Tox. Chem.* 8:843–844 (1989).
8. Holdren, J., Energy in Transition, *Sci. Am.* Sept:156–163 (1990).
9. Ehrlich, P. R. and A. H. Ehrlich, *Healing the Planet,* Menlo Park, CA: Addison-Wesley (1991).
10. Skinner, B. F., *Upon Further Reflection,* Englewood Cliffs, NJ: Prentice-Hall (1987).

Index

223

DATE DUE

JUL 0 9 1996			